Civil Engineering Technology 3

Civil Engineering Technology 3

B. G. Fletcher

Senior Lecturer in Civil Engineering Studies, Vauxhall College of
Building and Further Education

S. A. Lavan

Lecturer in Civil Engineering Studies, Herefordshire Technical College

Additional material by:

D. C. Wallis

Senior Lecturer in Civil Engineering Studies, Vauxhall College of
Building and Further Education

Illustrations prepared by:

Peter R. Bowyer

NEWNES-BUTTERWORTHS
TEC
TECHNICIAN SERIES

THE BUTTERWORTH GROUP

UNITED KINGDOM
Butterworth & Co (Publishers) Ltd
London: 88 Kingsway, WC2B 6AB

AUSTRALIA
Butterworths Pty Ltd
Sydney: 586 Pacific Highway, Chatswood, NSW 2067
Also at Melbourne, Brisbane, Adelaide and Perth

CANADA
Butterworth & Co (Canada) Ltd
Toronto: 2265 Midland Avenue, Scarborough,
Ontario M1P 4S1

NEW ZEALAND
Butterworths of New Zealand Ltd
Wellington: T & W Young Building,
77—85 Customhouse Quay, 1, CPO Box 472

SOUTH AFRICA
Butterworth & Co (South Africa) (Pty) Ltd
Durban: 152—154 Gale Street

USA
Butterworth (Publishers) Inc
Boston: 10 Tower Office Park, Woburn, Mass. 01801

First published 1980

© B. G. Fletcher and S. A. Lavan, 1980

British Library Cataloguing in Publication Data

Fletcher, B. G.
    Civil engineering technology 3.
    1. Civil engineering
    I. Title II. Lavan, S. A. III. Wallis, D. C.
    624   TA145   79-40954

ISBN 0 408 00426 6

Typeset by Butterworths Litho Preparation Department
Printed in England by Page Bros Ltd, Norwich, Norfolk

Preface

The aim of this book is to provide a background coverage to the TEC standard essential unit *Construction Technology III B* (U75/076). As the book is intended for students in the civil engineering discipline, we consider that the title *Civil Engineering Technology 3* is more appropriate.

Although it is recognised that at present there would seem to be no shortage of textbooks covering the principles of construction there appears to be a need for a single book that covers the requirements of a student at technician level studying civil engineering construction. We hope that this volume will meet these requirements. In this book we have used the term 'Topic Area' in place of the more usual 'Chapter' as this forms a recognisable link with the standard unit. The depth of treatment for each topic reflects the requirements of the unit weighting scheme as described by the TEC. Suggested further reading is included at the end of each topic area.

It is not the intention of this book to provide a comprehensive guide to all practices involved in civil engineering construction, as this would be impractical. Our aim has been to cover the work of the standard TEC unit U75/076, treating each topic to the level of understanding required of the unit.

It is assumed that students studying this unit would either have been exempt from, or have been successful in the TEC standard units *Construction Technology I* (U75/073) and *Construction Technology II* (U75/074). Where reference has been made to other TEC units a brief explanation has been included in the text of this book to integrate the two.

The authors should like to extend their thanks to David Wallis for assisting in the preparation of Topic Area E, Robert Rudd for reading the proofs and making helpful suggestions, Peter Bowyer who has so expertly prepared the illustrations and finally, but not least, to Mrs Susan McDonald who so patiently typed the manuscript.

B. G. Fletcher
S. A. Lavan

Acknowledgement is made to the Controller of H.M. Stationery Office for permission to reproduce Figures 117 and 118 and Table 12. Material from British Standards is reproduced by permission of the British Standards Institution, 2 Park Street, London W1A 2BS from whom complete copies can be obtained.

Contents

Topic Area A – General

1. NECESSITY AND SCOPE OF SOIL INVESTIGATIONS

The understanding of the physical makeup, engineering and chemical properties of the sub-soil strata of a site is of fundamental importance to the safe design of the structure. To obtain this information a soil investigation must be carried out.

For a full understanding of the requirements and techniques involved it is essential to refer to two codes of practice which are of particular relevance to this work:

(i) CP 2001: 1957. *Site Investigations*
(ii) BS 1377: 1975. *Methods of testing soils for Civil Engineering purposes.*

1.1 Scope of soil investigations
The term sub-soil is used to describe the layers of strata that lie between the top-soil (see Figure 1) and the bedrock. It is these layers of soil that in most cases support or surround (for example, in the case of tunnels) a structure.

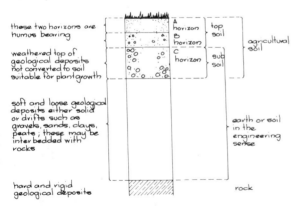

Figure 1   Relationship between engineering and non-engineering soils (*From CP 2001: 1957*)

The scope of a soil investigation in terms of both extent and detail will depend on several factors. These may in general terms be stated as:

(i) The type and importance of the structure.
(ii) The existence of any previous knowledge of the sub-soil.

(iii) The cost of the investigation relative to the cost of the proposed structure.

Assuming then that a site requires a fairly detailed report the soil investigation report should contain sufficiently detailed information on:

(i) The nature and thickness of the sub-soil strata.
(ii) The mechanical properties of each stratum or at least the strata and any underlying strata that a structure is to be supported on. Such information would include the density, natural moisture content (including any expected seasonal variations) compressibility, bearing capacity and shear strength.
(iii) The chemical and physical properties.
(iv) The natural ground water level of the site, again noting any seasonal variations.

This will enable conclusions to be drawn on the behaviour of the soil during excavation, construction and the working life of the structure. The report will also have a major influence on the type of foundation a structure will require.

1.2 Investigation methods
Before samples of the soil may be taken, access must be made to the various levels of soil strata. The choice of how this is done will depend on the nature of the ground (e.g. whether clay or running sand, etc.) the topography of the ground and the comparative cost of the available methods.

The actual methods most commonly used are:

(i)   Trial pits;
(ii) Borings;
(iii) Headings.

(i)   Trial pits

By digging a hole large enough for a person to work in, a section through the ground will be exposed, revealing the soil strata for examination and sampling. This method is normally used for shallow exploration up to depths of about 3.0 m. Depths greater than 6.0 m do not tend to be economic in comparison to the second method.

1

RECORD OF BORING

GROUND SURFACE LEVEL: + 42.00 O.D. (Newlyn)

TYPE OF BORING: Rotary drill

DIAMETER OF BORING: 150 mm

INCLINATION: 90°

CONTRACT:

FOR:

LOCATION: 468

BORING NO: 1

LINING TUBES: 150 mm to depth of 15 m

DATE STARTED: 23.7.79

DATE COMPLETED: 30.7.79

| Description of strada | Depth below Surface level (m) | | Thickness (m) | O.D. lever of lower contact | Samples | | | Ground water | Remarks |
	From	To			Type	No.	Level		
Top soil	0	1.0	1.0	+ 41.00					
Stiff boulder clay with stone matrix	1.0	6.0	6.0	+ 36.00	U	12	+ 40.00 to + 38.00	Struck ground water at 35.00	
					U	13	+ 37.00 to + 36.50		
Brown silty clay	6.0	8.0	2.0	+ 34.00	D	14	+ 35.00 to + 34.00		

Figure 2 A typical borehole log

(ii) Borings

As with a trial pit the purpose of a borehole is to reveal the make up of the underlying strata but, in this case, information is obtained by the collection of samples at suitable intervals as the borehole is sunk.

It is very important to keep a detailed record or log of the progress of the bore. Figure 2 shows the recommended layout of a borehole log as described in CP 2001: 1957.

Boring equipment

(a) Post-hole auger. A simple hand operated tool that is suitable for depths of up to approximately 6.0 m in soft strata, (which will stand unsupported) is shown in Figure 3. In some cases such as gravels or

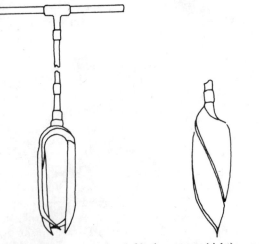

Figure 3 Post hole augers: (left) clay auger; (right) gravel auger

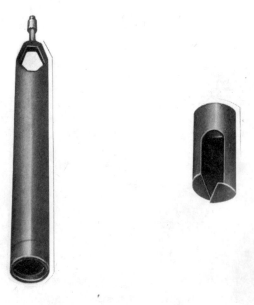

Figure 4 Shell with auger attachments

2

loose sands the bore may be lined but the placing of the lining may require mechanical assistance.

(b) Shell and auger boring. This tool can be hand operated in soft soils up to a depth of about 20 m with a diameter of 200 mm. For greater depths a mechanical tool would be employed. Casing (if required) are positioned by means of a 'monkey' suspended from the winch (see Figure 4).

(c) Percussion. As the name suggests the operation of a percussion borer is by repeated blows, breaking up the formation to sink the borehole. Water is added to the borehole as work proceeds, and the soil removed at intervals. Due to the action of the borer the collection of undisturbed samples must be carried out with care and for detailed investigations the rotary method may be more preferable. An advantage of the percussion borer is its rapidity of progress.

(d) Wash boring. With this method a tube is sunk by means of a strong jet of water. The jet of water disintegrates the soil and returns the disturbed soil by way of the returning current of water. Progress is made by either the tube sinking under its own weight or driven by a 'monkey'.

Figure 5 A typical mobile multipurpose drilling rig (Atlas Copco)

*(e) Rotary (*Figure 5*).* Two types of rotary drilling are used.

1. Mud-rotary drilling and,
2. Core drilling.

Mud-rotary drilling. In this system the bore is sunk by a drilling action with a rotating bit. As the hole is sunk a mud-laden fluid is injected into the borehole through the hollow rods connecting the bit to the rig.

In a manner similar to the wash boring technique the fluid carries the disturbed soil up the borehole to the surface. The mud-laden fluid also acts as a support to the sides of the hole so casing is not necessary. Samples are obtained by use of a core cutting tool.

Core drilling. Used in rock, the core drilling technique produces a continuous core of rock; the broken rock displaced by the core cutter is removed in a similar manner as the wash boring method.

(iii) Headings

In suitable topographical conditions the use of headings driven either horizontally or inclined may be used. It is, for example, a particularly useful method to explore steeply dipping strata. In tunnelling work it is normal practice to drive a pilot tunnel ahead of the main tunnel to explore the ground conditions, this is a form of the 'heading' technique.

Table 1 Soil investigation methods (*From CP 2001: 1957*)

Method/conditions	Soil	Rock
1. Trial pits	Open Timbered Piled Caissons	Open Timbered
2. Borings	Post-hole auger Shell and auger Wash boring Rotary	Percussion Rotary
3. Headings	Timbered Lined tunnels	Open Timbered

The choice of method may be summarised by reference to Table 1.

Borehole casings

In soft strata the borehole may collapse if the wall of the hole is not supported. The method most commonly used to prevent this happening is to line the hole with steel tubes, see Figure 6.

Figure 6 Typical borehole casing

High tensile steel tubes in lengths of 1.5 m to 3.0 m with a male and female screwed ends, are driven down just ahead of the boring tool by a drop hammer operated by the same winch as the tool. Care must be taken when obtaining undisturbed samples or carrying out penetration tests at the base of the hole, not to drive the casing too deeply, as this may affect the test results. The diameter of the casing should allow for a close fit with the boring tool.

Table 2 General guide to borehole diameters for varying depths

Total depth of borehole	Starting dia. of borehole	Final reduced dia. of borehole
Up to 20 m	150 mm	150 mm
20 m to 40 m	200 mm	150 mm
40 m to 60 m	250 mm	150 mm
60 m to 80 m	300 mm	150 mm

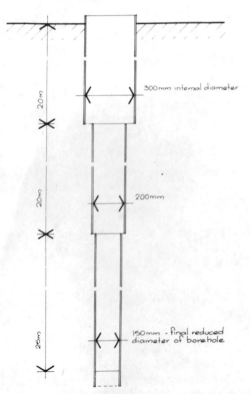

Figure 7 Reduction of casing diameter to facilitate withdrawal

To facilitate the withdrawal of the casing after the completion of the work in deep bore holes (i.e. over 20 m), the diameter should be reduced in stages. Suggested values for the reduction are set out in Table 2. To illustrate the procedure Figure 7 shows a typical lined borehole:

1.3 Soil samples

Two types of soil samples are used for soil investigation analysis;

(i) Undisturbed samples and
(ii) Disturbed samples.

(i) Undisturbed samples

Certain soil tests require the sample of soil used, to be in an undisturbed state. That is, the sample must retain the natural soil structure, moisture content, and void-ratio of the soil, from which it was taken.

Methods of obtaining undisturbed soil samples. Trial pits. Hand samples of clay or sand are easily obtained from trial pits. For clay a cube may be cut out with a sharp knife or a hand-corer of standard dimensions and known weight, see Figure 8. The cylindrical cutter is pushed into the clay, pulled out and weighed. As the weight and dimensions of the cutter are known, the weight of the sample can be obtained by weighing the cutter and sample after the removal of the soil. Hence the soil density can be calculated.

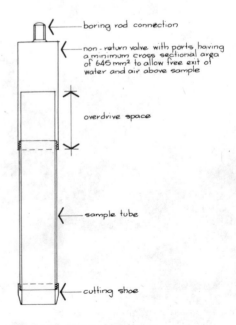

Figure 8 Standard hard corer

The cylindrical cutter can also be used in comparatively moist sands but it is very difficult to obtain an undisturbed sample by this method in any other type of soil. A useful method for sand or gravel is to push into the soil a sampling tube or box open at both ends, the surrounding soil is dug away and a trowel introduced underneath the tube or box. Level off the top and cover with a plate, the sample can then be lifted out.

It is important not to take a sample from soil that has been exposed to the direct rays of the sun, wind or rain.

Boreholes. The core cutters illustrated in Figures 9 and 10 which are used for borehole samples should be well oiled both inside and out before attaching to the boring rods. This is to reduce friction, as frictional resistance whilst using the cutter may disturb the natural soil structure of the sample. The tool is

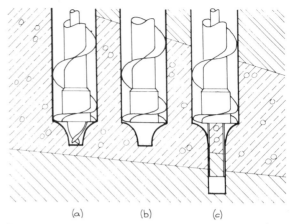

Figure 11   Hollow stem auger sampler: (a) auger flight and centre rod at desired depth; (b) centre rod and drill bit withdrawn; (c) sampling device introduced through hollow auger sections and sample withdrawn

Figure 9   U4 sampler

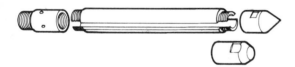

Figure 10   Split spoon sampler

lowered to the bottom of the borehole and forced into the clay by either jacking or by blows; jacking is to be preferred. The distance that the core is driven into the soil should be carefully monitored to prevent the soil compressing in the cylinder. The non-return valve at the top of the sampler allows air to escape as the soil is forced up the cylinder. The tool consists of a cutting shoe, a sampler tube and a driving head. These three sections are detachable, allowing the sampler tube to be used as a container in which the sample can be sent to the laboratory. Figure 11 illustrates an undisturbed sampling technique.

To obtain undisturbed samples in water laden soils a Bishop sand sampler can be used, see Figure 12. This is a compressed air sampler, the sample is removed from the ground into an air chamber and then lifted to the surface without contact with the water in the borehole.

When the undisturbed sample has been obtained it must be protected from damage and change in moisture content. As mentioned before, the core-cutter tube

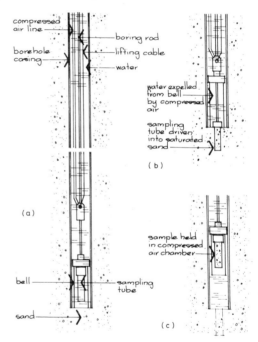

Figure 12   Bishop sand sampler: (a) sampler lowered into position; (b) sample collected; (c) sample withdrawn to surface

can be used as a casing for transit to the laboratory, but the two ends must be sealed. The ends of the tubes should first be covered by waxed paper and then several layers of molten paraffin wax applied. Screw-on lids are then put on and sealed with tape. If the core is removed from the sampler this must be done with great care to avoid remoulding of the sample. The sample should be wrapped in wax paper and coated in paraffin wax then placed in an air tight

```
No.
                          SAMPLE RECORD                          No.

      Location . . . . . . . . . . . . . . . .      Date. . . . . . . . . . . . . . . . . . .

      Boring No . . . . . . . . . . . .    O.D. of ground surface . . . . . . . . . . . .

      Position of sample, from. . . . . . . . . . . . . . . . . . . . . . . . . . . . . . .

      to. . . . . . . . . . . . . . . . . . . . . . . . . . . . . . . below ground surface.    No.

      Container No . . . . . . . .      Type of sample Disturbed/Undisturbed

                             REMARKS



                                                    Signed:           No.
```

Figure 13   Recommended sample label (*From CP 2001: 1957*)

jar. The space around the sample should be well packed with sawdust to prevent damage during transit.

(ii) Disturbed samples

Because of the method of extraction the natural moisture content and void ratio of a disturbed sample may vary from its *in situ* values, but the soil constituents will remain unaltered. Disturbed samples are used either for tests to supplement the information obtained from undisturbed samples or to determine the suitability of the soil for earthworks or stabilisation.

Such samples should be taken in duplicate, one for testing and one for the client, if required. For clays, at least 1.0 kg should be taken for each sample and placed in an air-tight jar. Sands and gravels should be taken in greater quantities. The recommended minimum amounts are

(a) medium-grained soils 5.0 kg
(b) coarse-grained soils 30.0 kg.

These samples can be stored in well tied polythene bags unless the moisture content is of importance then the sample should be placed immediately on recovery in an air-tight jar.

Because of the rapid variations that can take place in the properties of soils, especially clays, and as these changes may not be apparent by visual examination alone, samples should be taken frequently. It is recommended that if possible, a 100 mm dia core be taken when a major change in soil type is evident and at about 1.5 m intervals during borings through apparently homogeneous stratum.

All samples should be labelled immediately after being taken from the bore-hole or trial pit. Figure 13 shows the recommended form to be used. These labels are pre-printed in duplicate and bound in book form. The serial number on the right hand detachable portion, which is attached to the sample, is printed three times so that the chance of it being defaced is diminished.

To summarise, the type of method of sampling, and the type of sample obtained are shown in Table 3.

Table 3   Summary of sampling methods

	Type of sample	*Method of sampling*
Soils	Disturbed	Hand samples Auger samples (e.g. in clay) Shell samples (e.g. in sands)
	Undisturbed	Hand samples Core samples

6

1.4 Soil testing

It is frequently advantageous to determine the shear strength and density of the sub-soil *in situ* instead of obtaining samples from trial pits or boreholes for laboratory testing. Site testing is particularly useful in soft clays and sands where boring may result in some disturbance of the soil structure making the recovery of undisturbed samples very difficult. Some of the tests used are described below.

(i) Static penetration test

Sometimes referred to as the 'Dutch Cone' test due to its origin. The test consists of driving a rod with a point through a casing into the soil at a constant rate of loading. As the rod is eased there is no frictional resistance along the rod, therefore only the point of resistance is recorded, indicating the relative density along the penetrated depth. Only used in silty and soft strata conditions such as found in river estuaries.

(ii) Dynamic penetration test

Widly used in the UK and known as the Standard Penetration Test. In clays and sands a split-barrelled sampling tube of standard dimensions should be used (see Test 18 BS 1377: 1975) and in gravels a closed conical ended rod.

Before the test commences the base of the bore-hole should be relatively clean to avoid carrying out the test in partly disturbed conditions. The procedure is as follows.

The sampler is seated in the bore hole by being driven 150 mm into the soil by means of a 65 kg drop hammer with a fall of 760 mm, the number of blows required to obtain the 150 mm penetration being recorded. The sampler is then driven 305 mm or until 50 blows have been applied by the drop hammer, again with a free fall of 760 mm. The number of blows required to obtain each 76 mm of penetration is recorded. The total number of blows required for the full depth of penetration, i.e. 305 mm is termed the 'penetration resistance' N. If, after 50 blows, the penetration is less than 305 mm, a record of the number of blows and depth of penetration is made.

Table 4 Guide to density of soil according to number of blows

No. of blows/305 mm	Relative density
0 to 4	Very loose
4 to 10	Loose
10 to 30	Medium
30 to 50	Dense
Over 50	Very dense

Terzaghi and Peck recommend the following figures shown in Table 4 or the number of blows for full penetration as a guide to the *in situ* density.

Conclusions on the bearing capacity can be drawn from their results.

(iii) The California bearing ratio (CBR)

In the design of runways and roads the strength of the sub-grade (see E. 18.2) is a principal factor in determining the thickness of the pavement. The strength of the sub-grade is assessed on the California Bearing Ratio of the soil.

For a general description of the apparatus and method of use, see objective E. 18.3. For a detailed description reference should be made to BS 1377: 1975 Test 16 and for guidance on the application of the results, Road Research Laboratories publication *'Road Note 29'* should be consulted.

(iv) Vane tests

In soft clays it can be difficult to obtain reliable undisturbed samples to test for shear strength in a

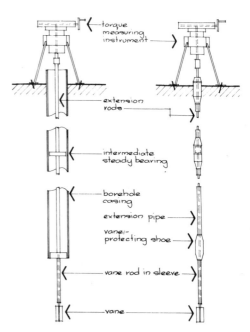

Figure 14 Vane test apparatus: (left) borehole type; (right) penetration type

laboratory. To overcome this problem, an *in situ* method of testing the shear strength of soft clays and silts is carried out. The test is known as the 'vane test' and can either be hand-worked or mechanically driven.

7

Two types of apparatus are used, they are:
(a) Borehole (Figure 14)
(b) Penetration.

Basically a vane of cruciform section is driven into the soil and subjected to a torque force. The measure of shear strength is the relationship of the torque required to cause a cylindrical surface of rupture and the diameter and height of the vane.

1.5 Grading test

The purpose of a grading test is to determine the particle size distribution of a soil. Soils, are for the purpose of particle size classification grouped as follows:

(a) *Fine-grained soils.* Soils containing not less than 90% passing a 2.36 mm BS sieve.
(b) *Medium-grained soils.* Soils containing not less than 90% passing a 20 mm BS sieve.
(c) *Coarse-grained soils.* Soils containing not less than 90% passing a 40 mm BS sieve.

Group (a) are further classified as shown in Table 5.
The analysis for the above classifications is carried out in two stages:

1. The separation of coarser fractions by sieving on a series of BS sieves.
2. The determination of the proportions of the finer particles by a sedimentation process, generally known as 'wet analysis'.

Sieving

The sample of soil is oven dried for 24 hrs at 105–110°C. The sample is then weighed and sifted through a series of BS sieves, (sizes 10 mm, 5 mm, 2.36, 1.18 mm, 600, 300, 212, 150 and 74 μm) the amount of soil retained in each sieve is then weighed. To illustrate the recording and presentation of the results an example of a sieve analysis is given in Table 6.

Table 6   Typical result of grading test

B.S. sieve size	Weight of material retained on sieve	Weight of material passing each sieve	% of material passing sieve
10 mm	6.47 g	168.13 g	96.3
5 mm	12.57 g	155.56 g	89.1
2.36 mm	19.73 g	135.83 g	77.8
1.18 mm	21.82 g	114.01 g	65.3
600 μm	35.61 g	78.40 g	44.9
300 μm	37.54 g	40.86 g	23.4
212 μm	9.44 g	31.42 g	18.0
150 μm	5.75 g	25.67 g	14.7
75 μm	9.96 g	15.71 g	9.0
< 75 μm	15.71 g		
Total wt.	Σ 174.60 g		

A grading curve is then drawn by plotting the % of material passing each sieve against BS sieve sizes used. This is illustrated in Figure 15.

Wet sieving

The sample is prepared as before by oven drying and weighed. The sample is then sifted through a 20 mm BS sieve, the remainder of the original sample that passes through the 10 mm sieve is then reduced to a convenient amount (for maximum sieve loads see Form F, BS 1377: 1975 through a riffle box, the weight of the sample is then recorded.

Table 5   Classification of fine-grained soils

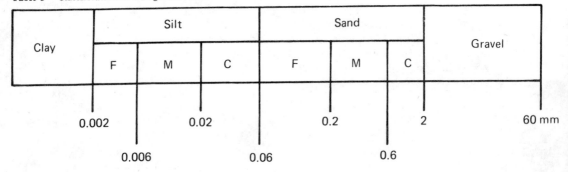

F: Fine.   M: Medium.   C: Coarse.

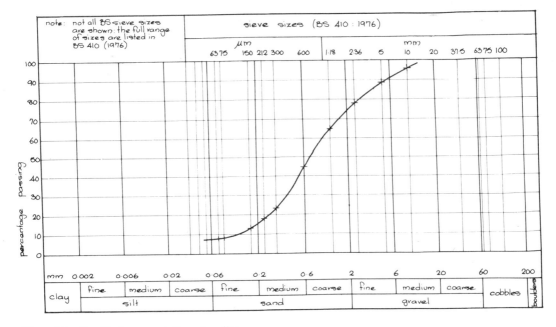

Figure 15 Graphical representation of grading

The sample is placed on a tray or in a bucket and covered with water. A solution of sodium hexametaphosphate is added at a rate of 2 g per one litre of water, the whole being well stirred. The sample is then left to stand for one hour in the solution but stirred frequently.

After standing for one hour the sample is washed through a 5 mm BS sieve into a tray the fraction retained on the 5 mm sieve is to be kept for oven drying and weighing. The remainder of the sample is then washed through the rest of sieves made up of the following BS sieves.

2.36 mm; 1.18 mm and 75 μm

The fraction of the sample that passes the 75 μm sieve is then discarded and the process repeated until the water after washing through the sieves is virtually clear.

Finally the remaining fractions of the sample are placed in separate trays and oven dried at 105–110°C for 24 hours. The dried samples are then sieved in the same manner as for the sieving test and the results recorded and presented as previously illustrated.

Sulphate content of soil and soil water

The presence of soluble sulphates in soils can cause the breakdown of Portland cement concretes in contact with water in such soils, due to the formation of calcium sulphurluminate. Due to the expansion of volume accompanying the formation of the compound internal stresses are set up sufficient to disrupt the cement matrix and cause mechanical failure of the material as a whole.

Expressed in terms of sulphate trioxide the presence of less than 0.2% SO_3 in the soil or soil water is unlikely to cause any problems but should the content of SO_3 be 0.5% or above there is a serious risk of deterioration of the concrete. Soluble sulphates are also a major factor in the corrosion of metal pipes laid in waterlogged clay soils.

The procedure and calculations for the determination of water soluble sulphate content are fully described in BS 1377 (1975). Test 9 describes total sulphate content of soil and test 10 describes the determination of the sulphate content of ground water and aqueous soil extracts. When obtaining results from either test it should be remembered that the moisture content of soils will fluctuate according to the time of the year, and hence the sulphate content.

1.6 Identification of soils

The knowledge required to interpret a borehole log is beyond the scope of this book, as it requires a good understanding of soil mechanics. There are certain ways of obtaining a general impression of the expected performance of soils. Tables 7, 8 and 9 will be of assistance.

Table 7 Approximate densities of soils

Material	*Density kg/m³*	*Material*	*Density kg/m³*
Chalk	2000	Gravel, with sand	1700
Clay, dry	1600	Loam, dry	1600
Clay, wet	2000	Loam, wet	2100
Earth, dry loose	1300	Mud	1600
Earth, dry rammed	1750	Pebbles	1750
Earth, damp	1600	Sand, dry	1500
Earth, very compact	1850	Sand, damp	1750
Earth, wet rammed	2000	Sand, wet	1900
Gravel	1750	Stones, broken	1600

Table 8 Maximum safe bearing capacities of soils

Material	*MSBC kN/m²*	*Material*	*MSBC kN/m²*
Limestone	1280	Clay, soft	55–110
Slate	3800	Clay, very soft	0– 55
Shale, hard	2140	Gravel, compact	430–640
Sandstone, soft	2140	Gravel, loose	215–430
Clay/shale	1070	Sand, compact	215–430
Boulder clay,	430–640	Sand, loose	110–215
Clay, stiff	215–430		
Clay, firm	110–215		

Table 9 General basis for field identification and classification of soils (*From CP 2001: 1957*)

	Size and nature of particles		
	Principal soil types		
	Types	*Field identification*	*Composite types*
Coarse-grained non-cohesive	Boulders Cobbles	*Larger than* 200 mm dia. *Mostly between* 200 and 75 mm	Boulder Gravels
Coarse-grained non-cohesive	Gravels	Mostly between 75 mm and 2.36 mm BS sieve	Hoggin Sandy gravels
Coarse-grained non-cohesive	Sands, uniform and graded	Composed of particles mostly between 2.36 mm and 75 μm BS sieves, and visible to the naked eye. Very little or no cohesion when dry. Sands may be classified as uniform or well graded according to the distribution of particle size. Uniform sands may be divided into coarse sands between 2.36 mm and 600 μm BS sieves, medium sands between 600 μm and 212 μm BS sieves and fine sands between 212 μm and 75 μm BS sieves.	Silty sands Micaceous sands Lateritic sands Clayey sands
Fine-grained cohesive	Silts, low plasticity	Particles mostly passing 75 μm BS sieve. Particles mostly invisible or barely visible to the naked eye. Some plasticity and exhibits marked dilatancy. Dries moderately quickly and can be dusted off the fingers. Dry lumps possess cohesion but can be powdered easily in the fingers.	Loams Clayey silts Organic silts
Fine-grained cohesive	Clays: Medium plasticity	Dry lumps can be broken but not powdered. They also disintegrate under water. Smooth touch and plastic, no dilatancy. Sticks to the fingers and dries clowly. Shrinks appreciably on drying, usually showing cracks.	Boulder clays Sandy clay Silty clays Maris Organic clays
Fine-grained cohesive	High plasticity	Lean and fat clays show those properties to a moderate and high degree respectively.	Lateritic clays
Organic	Peats	Fibrous organic material, usually brown or black in colour.	Sandy, silty or clayey peats

Table 9 *(cont'd)*

| Strength and structural characteristics | | | |
| Strength | | Structure | |
Term	Field test	Term	Field identification
Loose	Can be excavated with spade. 50 mm wooden peg can be easily driven.	Homogeneous	Deposits consisting essentially of one type.
Compact	Requires pick for excavation. 50 mm wooden peg hard to drive more than a few centimetres.		
Slightly	Visual examination. Pick removes soil in lumps which can be abraded with thumb.	Stratified	Alternating layers of varying types.
Soft	Easily moulded in the fingers.	Homogeneous	Deposits consisting essentially of one type.
Firm	Can be moulded by strong pressure in the fingers.	Stratified	Alternating layers of varying types.
Very soft	Exudes between fingers when squeezed in fist.	Fissured	Breaks into polyhedral fragments along fissure planes.
Soft	Easily moulded in fingers	Intact	No fissures.
Firm	Can be moulded by strong pressure in the fingers.	Homogeneous Stratified	Deposits consisting essentially of one type. Alternating layers of varying types. If layers are thin the soil may be described as laminated.
Stiff	Cannot be moulded in fingers.	Weathered	Usually exhibits crumb or columnar structure.
Hard	Brittle or very tough.		
Firm	Fibres compressed together.		
Spongy	Very compressible and open structure.		

1.7 Site report

So far we have been concerned with the practices of soil investigation. This will form only part of the information that will be required by the managing consultants and contractor, before a start can be made on the construction of the intended structure. The complete report, that will also contain the soil investigation report, is referred to as the 'Site Report'.

It is recommended that a comprehensive site report with information on the following items, if and when relevant, should be supplied.

(i) General information

(a) Ordnance and geological maps.
(b) General survey; i.e. location of site, corrections on existing OS maps, property lines etc.
(c) Approaches and access.
(d) Restrictions: i.e. legal restrictions or existing structures or works beneath the site, rights of way etc.

(e) Drainage and sewage.
(f) Supply of services; i.e. electricity, gas, water etc.

(ii) Special information required for design and construction

(a) Detailed survey.
(b) Present and past uses of the site.
(c) Soil investigation report.
(d) Hydrological report in the case of river and sea works.
(e) Climatic conditions.
(f) Sources of local materials for construction.
(g) Disposal of waste materials i.e. distance to and availability of spoil tips.

Many contractors use pre-printed forms to be filled in as the information is acquired, this will help to avoid omissions. These could prove very expensive at a later date. Part of a typical site investigation report form is shown in Figure 16.

7 DEMOLITION OF EXISTING BUILDINGS	If any shoring and/or underpinning required to adjacent buildings and liability for their condition		

8	Details	Load Capacity	Width
ACCESS TO SITE	Existing roads on site		
	Existing roads adjacent to site		
	Temporary roads required		
	Access difficulties		

9	Strata expected to be found during excavation					
GROUND CONDITIONS (STATE MAX. DEPTH AND LOCATION OF STARRED ITEMS)	Boreholes*	Boreholes Existing YES/NO			Boreholes Required YES/NO	
	Effect of Adverse Climatic Conditions on Site Operations					
	Water	Surface YES/NO	Depth of Table (m)	Tidal YES/NO	Springs* Yes/No (If yes give number)	Pumping required YES/NO
	Any existing foundations, roots, rock or any other material likely to be found in ground					
	Stability of excavations					
	Timbering required					
	Any other ground details*					

Figure 16 Extract from site investigation report form

12

2. ORGANISING, LAYING OUT AND ARRANGING ACCESS TO SITES

If a contract is to be carried out in an efficient manner, an essential element of this goal is the planning of the site layout, before the contract is started. No two sites are the same and 'perfect' solutions do not exist, hence the site plan adopted for a contract will be an optimum solution to a problem that has many variables to be taken into account. The aim of a site planning exercise is to produce a layout that is logical, orderly and above all practical.

2.1 Site layout

In general terms the main items that must be considered in laying out a new site may be listed as follows:

(i)   Access to the site and on site roads.
(ii) Storage of materials.
(iii) Plant requirements and movement of plant.
(iv) Temporary buildings for the contractors team and other persons resident on the site.
(v)   Temporary services.
(vi) Fencing and hoardings.

2.2 Items to be considered

(i)   Access to the site and on site roads

Access to the site would normally be described in the contract. If the site has more than one point of access each entrance should be identified by either a letter or number and instructions clearly displayed as to where visitors and material deliveries should report on arrival. To facilitate on-off movement of site traffic a system of exit and entrance only gates can be of advantage.

The movement of plant on a site must be planned for the efficient and economic operation of the machines. This is particularly so, when siting haulage roads to spoil tips. The decision on what type of roads are to be used, e.g. rough access, waterbound or bitumen sprayed, will depend upon the type of plant that is to be used, the ground conditions of the site and the cost effectiveness.

For example if the first two points are considered; tracked vehicles fitted with grips cannot pass continuously over waterbound or bitumen-sprayed roads, as the grips on the tracks will destroy the surface. Heavy plant on flat tracks, such as diggers, cannot pass over soft ground. Scrapers can pass over soft or hard ground without seriously damaging the surface; dump trucks apply high concentrated wheel loads and therefore cannot pass over soft ground and so on.

Generally rough access roads are used by on-site traffic and waterbound or bitumen-sprayed roads for on-off site traffic. These roads should be maintained to prevent the breaking up of the haunches and ditches and potholes occurring that could damage wheeled vehicles.

The layout of site roads should provide for a smooth movement of all site traffic with an economy of distance. Where off site vehicles cross over or use on site vehicle roads, warning signs must be posted or be supervised for reasons of safety.

(ii)   Storage of materials

The aggregate and cement stores should be adjacent to the mixing plant, allowing sufficient access for delivery trucks to unload at the batching area.

Specific objective C9.2 describes the method of storing cement and aggregates and reference should be made to this objective.

All materials should be stored in an orderly fashion. A great deal of wastage can occur if the storage methods are lax. As a general rule the storage of materials should be so positioned as to avoid double handling and awkward access for delivery trucks.

(iii)   Plant requirements and movement of plant

For the choice and use of excavating plant see specific objective B3.1.

In general terms the location and working procedure of plant should aim to minimise the repositioning of the machine; and maximise the coverage of area that the plant is required to operate over. This is particularly so for cranes.

Tower cranes, unless mounted on rails, derricks and cableways should not have to be repositioned during a contract, as the process is costly and, of course, during repositioning work the crane will be out of action. Excavating plant, particularly on extended sites such as roads and airfields should 'work as they go' for example a scraper should start its run directed towards the spoil tip. Concreting plant should be positioned to give minimum delivery distance to the areas of the site that require the bulk of the concrete output.

(iv)   Temporary buildings

The contractors' main offices and the general storage buildings for site equipment should be located near the main entrance to the site. As all materials delivered to the site must be checked on arrival, the store-keeper's office must be at the main entrance. Care must be taken to avoid obstructing the entrance to other vehicles while checking is carried out.

For good communications it is wise to establish the resident engineers' offices close to the contractors' offices. If room allows, sub-contractors' offices could also be in this location. This will lead to an economy in providing such services as telephones, heating, lighting, cleaning, etc.

If this layout is followed an administrative area will be logically situated around the main entrance to the site thereby decreasing the number of persons walking or driving around the construction areas. A final point is that where feasible the contractor's and resident engineer's offices should be positioned to allow an overall view of the site from their windows including the main entrance point.

(v) Temporary services

It is the contractor's duty to arrange the supply of all temporary services such as telephones, electricity and sanitation. Consideration must also be given to the diversion, due to the contract, of any existing services. It is usual for the engineer to supply the contractor with sufficient detailed information to locate the services in question and contact any interested bodies such as the Post Office, water and electricity authorities etc. It will also be necessary to arrange who will be responsible for the diverting of any services, i.e. the contractor or authority concerned.

(vi) Fencing and hoardings

Fencing and hoardings serve two purposes:

(a) To control the delivery and removal of materials from the site by controlled entry and exit points,
(b) To prevent unauthorised persons entering the site. In rural areas this would also apply to livestock.

It is the contractor's duty to erect and maintain the boundary fencing or hoardings. The type chosen will depend on the contractor's preference and the location of the site. On certain sites, particularly in cities, extra security may be required. In such cases professional security firms may be hired by the contractors.

2.3 Site planning

As stated previously, before a contractor can start work he must plan how the site is to be laid out. Planning the order of construction is equally important. Without such a plan the contract would

soon run into difficulties and is therefore considered a standard procedure.

The complexity of tasks involved on any one civil engineering contract will of course vary according to the type of work to be carried out, from the simplest single pipe-laying contract for a local authority to major schemes such as a power station or a dry dock complex. With any contract, however, the contractor must be able to state:

(a) How long it will take to complete;
(b) How many workers will be required at any time during the contract;
(c) The materials required and in what quantities they will be needed at any time in the contract.

When the contractor has decided upon the order of construction he will submit his scheme to the consultant engineers for approval. If the scheme is approved, the contractor will then start on a detailed plan of the work to be carried out. There are several methods used for planning, the main difference between them being the level of sophistication. The basic requirement of any plan, and its presentation, is that it will show in a manner that is clear to all who have to read it, the expected starting and completion dates of the individual jobs that make up the project. Linked with this is the number of skilled and unskilled operatives required and the quantity of materials to be used. This last point is very important as the late delivery of materials can hold up the contract and this can prove to be very expensive for the contractor. By planning his material requirements beforehand the contractor can obtain delivery dates and if necessary alter his plan to fit the available delivery periods.

The plan is usually referred to as a 'programme of work' and is normally presented in the form of a bar-chart. At first appearance a bar-chart may seem simple enough, but one should not be deceived. A great deal of work and experience must go into it if it is to be a realistic programme that can be adhered to during the contract.

To avoid producing a bar-chart programme that is difficult to read because of the amount of information it contains, it is more usual to produce several bar-charts. This is normally done by having a bar-chart that will show the periods of time (and in some cases the materials required) for the major items of work for the complete contract. Then, augmenting this overall bar-chart with weekly or monthly bar-charts showing in greater detail the work to be carried out in the particular week or month. Each bar-chart should be marked up as the work proceeds so that the progress of the contract can be assessed by referring to the bar-charts. Figure 17 shows a typical bar-chart.

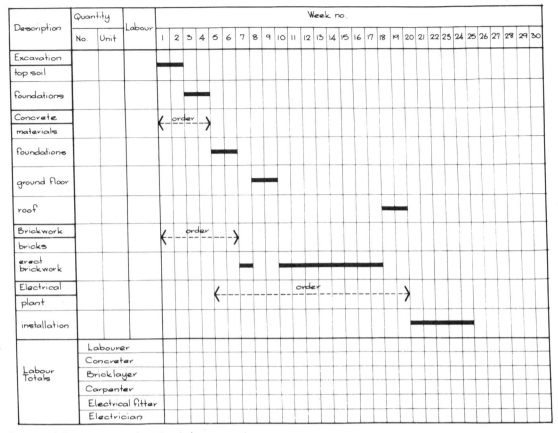

note : 1) this chart is only intended as a guide
2) the quantities of material and labour content have been left out, as these will depend on the size of the contract, length of programme, etc

Figure 17   Part of bar chart for electricity substation

2.4 Site personnel

To obtain a clear impression of the personnel and their tasks involved on a civil engineering contract it is necessary to distinguish between the contractor's and engineer's staff and employees.

When a contract is to be commissioned the client will appoint a firm of consultant civil engineers who will design the project and organise the selection of the main contractor; this may be done by open or selected tendering. It will also be the responsibility of the consultant to supervise the contract on behalf of the client.

This in simple terms, is the usual method of carrying out a civil engineering project, there are, of course, variations but this particular topic lies beyond the scope of this book.

For a typical contract the following personnel and their relative positions of responsibility may be set out as shown in Figure 18.

2.5 The contractor's site team

The key posts for a typical contract are the:

(i)   Agent.
(ii) Site engineer.
(iii) Plant engineer.
(iv) General foreman.
(v)   Office manager.

The responsibilities of the five posts are briefly described in the following paragraphs.

(i)   The agent. The agent is responsible for the overall running of the contract and will have the authority to hire men and plant, purchase materials and employ sub-contractors. He must of course, have an extensive knowledge of civil engineering techniques, but need not necessarily be a chartered civil engineer.

Many agents have risen from the ranks of the craftsman but as the techniques of construction have

1

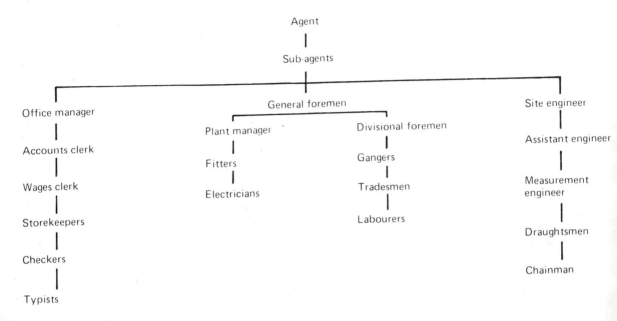

Figure 18 The contractor's site team

become more sophisticated many younger agents today have academic engineering qualifications.

(ii) The site engineer. The site engineer is responsible for the setting out of work. This will include taking site levels, lining in and levelling constructional work, planning and designing temporary works such as access roads, concrete batching plant; foundations and dealing with power and water supplies and the drainage of the site. The engineer will also be required to act as adviser to the site agent and keep progress and quality records of the work.

(iii) The plant manager. The plant manager's responsibility to keep all the mechanical plant in working order and to have it available as required by the programme of construction.

(iv) The general foreman. The general foreman's task is to keep the contract running to programme. He will issue the trades foreman the detailed instructions of work to be carried out and if necessary be able to show how it is to be done. To many people the foreman is the lynch-pin of the contractor's team.

(v) The office manager. The office manager is responsible for the administration work such as issuing orders for materials making up pay sheets, receiving and checking accounts.

In general many jobs on the site will overlap and chains or organisation will vary according to the practices of the contractor concerned and the type and size of contract.

(vi) The contract manager. On large contracts the contractor will appoint a contract manager who working from the head office will be concerned with the management rather than technical aspects of the contract. The manager in turn will be directly responsible to a director of the company.

2.6 The resident engineer's team

The consultant engineers will appoint a resident engineer (not to be confused with the contractor's site engineer) who, according to the size of the contract, will have a number of assistants. Such a team, for a medium project, might be as shown in Figure 19.

The resident engineer's main responsibility is to see that the construction of the project is in accordance with the designs and drawings issued by the consultant engineer and any specialist consultant (e.g. mechanical or electrical engineers) employed on the project. This will include checking that material specifications, workmanship, levels and setting out are as required, to measure completed work for the purposes of interim payment and to calculate such payments.

To assist him in his work the resident engineer will have a staff of assistant engineers and inspectors (in building contracts the clerk of works is the equivalent post). The number depending on the size and nature of the contract.

It is an important part of the resident engineer's duties to keep records of the work carried out. These records should enable a clear view of the progress of

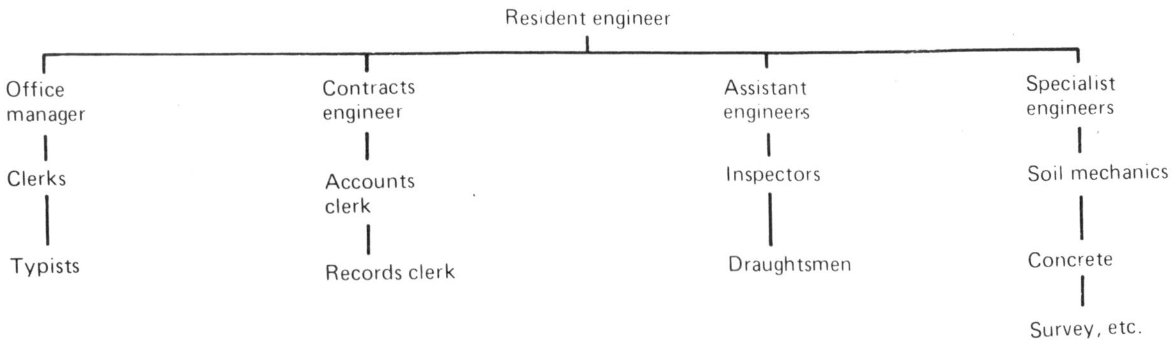

```
                        Resident engineer
                              |
 ┌──────────────┬──────────────────┬──────────────────┐
Office          Contracts          Assistant          Specialist
manager         engineer           engineers          engineers
 |              |                  |                  |
Clerks          Accounts           Inspectors         Soil mechanics
                clerk
 |              |                  |                  |
Typists         Records clerk      Draughtsmen        Concrete
                                                      |
                                                      Survey, etc.
```

Figure 19   The resident engineer's site team

the contract at any time to be obtained and form the basis on which, as mentioned above, interim payments will be made to the contractor. The engineer should also record the quality and performance of the materials used and the actual dimensions etc., of the completed work in the form of 'as built' drawings.

2.7 The Construction Regulations 1961

A contractor, as does any employer in the UK, has a legal obligation to ensure the health, safety and welfare of his employees at their place of work. This duty is enshrined in four sets of Regulations under the Construction Regulations 1961. They are:

1. Construction (General Provisions) Regulations 1966.
2. Construction (Lifting Operations) Regulations 1961.
3. Construction (Working Places) Regulations 1966.
4. Construction (Health and Welfare) Regulations 1961.

These Regulations apply to any type of building work and most types of civil engineering and construction work.

1. General provisions

This Regulation provides for safe working in the following areas:

(i)   Excavation, shafts and tunnels.
(ii) Cofferdams and caissons.
(iii) Explosives.
(iv) Dangerous or unhealthy atmospheres.
(v)   Work in or adjacent to water.
(vi) Site transport.
(vii) Demolition.

2. Lifting operations

This Regulation covers the use and maintenance (by required regular inspections) of lifting appliances, (i.e. cranes and derricks) chains, ropes, lifting tackle and hoists. It also covers the carriage of persons and the secureness of loads.

3. Working places

This Regulation provides that a safe route must be provided to and from every place of work on the site and that every place of work must also be safe. Any work that cannot be done from the ground or part of the structure msut have scaffolding erected for working from. It is this Regulation that covers the requirements for scaffold structures.

4. Health and Welfare

This Regulation, since its introduction, has had certain parts amended by the introduction of the Construction (Health and Welfare) (Amendment) Regulations 1974. The Regulation provides for the general health and welfare of construction site operatives and includes the provision of first-aid boxes and facilities for dealing with accidents, shelter, mess-rooms, washing and drying out facilities and sanitary arrangements.

It is planned that with the introduction of the Health and Safety at Work Act 1974 all existing safety legislation will eventually be replaced by this Act. Under this Act every employer of more than five employees must prepare a written statement of the companies health and safety policy. There are other acts and regulations which relate to certain aspects of civil engineering work such as the Mines and Quarries Act 1954; Diving Operations Special Regulations 1960, etc.

Under the Construction Regulations an employer who employs more than twenty persons must appoint a safety officer. This position may not necessarily be a full time one, but whoever is appointed must be given sufficient time from any other duties he has, to carry out his safety duties with reasonable efficiency. The name of the safety officer must be entered on the copy of the Regulations that the employer is required to display on site, normally in the contractor's office.

Further reading
Readers who require a more in depth knowledge of some of the objectives in Topic Area A may find the following reading list useful.

A.1. Capper, F. Leonard and Cassie, W. Fisher, *The mechanics of engineering soils*, E. and F. N. Spon Ltd. (1976).

A.2. Edmeades, Dennis, *The construction site*, Construction Management Series, Estate Gazette Ltd. (1972).

Twort, A. C., *Civil engineering supervision and management*, Edward Arnold (1972).

Construction safety, National Federation of Building Trades Employers (1975).

Topic Area B – Substructure

LARGE SCALE EARTH MOVEMENTS – TECHNIQUES, PROCEDURES AND PLANT

3.1 Choice of earth moving plant

On large civil engineering contracts the value of the earthworks as a percentage of the contract price can be as high as 60–70%; the final decision on the type of plant to be used can thus have an important influence on the profitability of the contract.

The factors to be considered can be wide and varied according to the type of site and contract, but in general terms the decisions taken must aim to achieve an optimum economy of working. General factors that would influence the choice may be listed as follows:

(i)   Location of site.
(ii)   Ground conditions of site.
(iii)   Time allowed to complete the work.
(iv)   Method of excavations to be employed.
(v)   Nature of material to be excavated.
(vi)   Haulage distance to tip.
(vii) Output of individual machines.
(viii) Available space for manouvering the machine.

These factors will form the basis of the decision on the number of machines, type and size and also the period of time the machines will be required on the site.

3.2 Types of excavating plant

(i) Bulldozers

The bulldozer or dozer is a track-mounted machine with a front blade. It is used for cutting and grading, assisting scrapers, clearance work and, when fitted with a ripper, breaking up rock beds. Every large site requires at least one large dozer to complete the necessary complement of earthmoving plant, see Figure 20.

Being track mounted, the weight of the machine is spread over a large area enabling it to work in all types of ground conditions whilst maintaining a high tractive power with little risk of becoming 'bogged down' in poor ground. Tracked vehicles are not allowed on metalled roads as they would rapidly break up the surface. Transportation to and from the site must be by a low-loading trailer. This applies to all track-mounted machines.

Figure 20 Bulldozer
(Caterpillar Tractor Co.)

(ii) Scrapers

This is a specialised excavator used for surface excavations over large areas, an approximate figure for the minimum length of cut that would allow the economic use of a scraper is about 70 m. The scraper can either be towed by a bulldozer or have its own power unit. The general principle of the machine is a large open-fronted bowl fitted with a retractable bottom cutting edge, which is drawn along and cuts (or scrapes) the material into the bowl.

A maximum depth of cut, in suitable ground conditions, is about 300 mm. When the bowl is full, or the cut completed, the cutting edge is raised clear of the ground and the scraper moves off to the dumping area.

The haulage distance is an important factor when deciding which type and size of scraper to be used. Bulldozer towed scrapers, because of their low speed, should be limited to 300 m hauls. Motorised scrapers, which have a much greater speed, can be used on hauls of up to two miles. In very difficult conditions both towed and motorised scrapers are sometimes assisted by a bulldozer pushing from the rear. This is particularly so, even in good conditions, for the final cut of fill. Figures 21 and 22 show typical examples.

The capacity of scraper bowls range from 7.5 m^3 to 46 m^3 on single bowl units and on twin bowl units may go up to 90–100 m^3 capacity.

(iii) Excavators/loaders

These are high output machines that can excavate, load and grade in all materials, including rock if it has been previously broken out with a ripper. The shovel can either be track or wheel mounted according to ground conditions and is usually hydraulically operated. In poor ground conditions the track-mounted unit has the advantage of being able to remain still while loading, as it can rotate on its tracks; this can prevent the problem of cutting up the formation.

Figure 21   Tandem-power scraper (Caterpillar Tractor Co.)

Figure 22   Bulldozer acting as a pusher on back of scraper (Caterpillar Tractor Co.)

The loader can excavate and load clean and accurately leaving a clean formation. One disadvantage is that it has a limited capacity to excavate below its standing level. To increase the manouverability of wheel mounted units articulated loaders are available. Shovel capacities range from 0.75 m³ to 6 m³. The size of bucket chosen will depend on the number and capacity of dump trunks available to cart away the material. This machine is illustrated in Figures 23 and 24.

On the smaller machines a hydraulic hoe (see 3.2 (*vi*)) is sometimes attached to increase the versatility of the machine.

(iv) Dragline

The dragline has a wide range of excavating abilities including trench work and cutting and grading of slopes. A track-mounted machine with a crane jib from which the bucket is suspended, it is capable of excavating below its standing level and off loading above or below its own level.

As with the scraper, its name suggests its method of working. The bucket is dragged along the ground, by cables, the teeth on the bucket digging in and the fill pushed up into the bucket. A disadvantage of the dragline is that loading the fill into trucks is difficult to do accurately, also the formation level, because of the teeth on the bucket, tends to be raked up. The dragline is illustrated in Figure 25.

The dragline can work in most materials as long as the teeth on the bucket can get a grip. It is used

Figure 23 Articulated wheel excavator/loader loading an off-highway dump truck (Caterpillar Tractor Co.)

Figure 24 Track mounted excavator/loader (Caterpillar Tractor Co.)

Figure 25 Dragline
(Ransomes & Rapier Ltd.)

Figure 26   Single chain ring discharge grab
(Butters Cranes Ltd.)

Figure 27a   Wheel-mounted hydraulic hoe with clam shell
attachment
(Akerman)

Figure 27b   Track-mounted hydraulic hoe excavator
(Akerman)

extensively for excavating in river beds and clearing channels, sometimes working off a pontoon.

(v) The grab

This machine, which can be either track or wheel mounted, has the ability to excavate deep trenches up to 30 m, piles in loose soils and underwater (see Figure 26). Most machines these days operate the clamshell bucket hydraulically although cable operated machines are available. For deep excavation work an extension rod can be fitted to the bucket. To assist in the accuracy of line when, for example excavating for piles or diaphragm walls, the arm of the bucket may be lowered down a frame held by a crane. The frame can be set vertically or on an incline giving the machines a high degree of accuracy when precise excavation is required.

(vi) Hydraulic hoe

This machine (also called a back-acter) with its hydraulically-operated hoe bucket, is universally employed for trenching in all types of ground excluding continuous rock, see Figures 27a and b. The hoe can be fitted with a rock breaking point and used to break out weakly cemented rock. The hoe can either be mounted on a wheel or tracked unit.

(vii) Continuous bucket trencher

For light soils without rocks, roots or service pipes, the continuous bucket trencher is a most efficient excavator. The buckets are carried either on a paddle

Figure 28 Continuous bucket trencher (Barber–Greene Ltd.)

wheel or continuous chains, the supporting boom can be adjusted to give an accurate depth of cut up to 4 m deep, with a width of dig, according to the bucket size from 125 mm up to 1.75 m.

In order to be used economically it should cut at least one mile of trench a day. It is extensively used for long pipe-laying contracts such as for the North Sea gas pipelines in the UK. Figure 28 illustrates the machine.

3.3 Excavation support work

It is not the intention of this section to describe in detail the different methods available of supporting deep excavations as the method used for any one case will depend on a number of factors. These factors would include the soil conditions, the location of the work, the type of plant to be used, the length of time the excavation will be open and the reason for the excavation.

At this level of study the reader should become aware of the basic requirements of excavation support work and gain an appreciation of the general principles involved.

Part IV of the Construction (General Provisions) Regulations requires that an adequate supply of timber or other support material be provided and used in excavations of more than 1.25 m in depth. This requirement is to prevent danger to any person working within an excavation from a fall or dislodgement of earth, rock or any other material forming the sides of the excavation. Further requirements of this Regulation are that:

(i)   A competent person should inspect the excavation at the start of each day's work or at the beginning of esch shift.

(ii)   A thorough examination be carried out after the use of explosive charges, damage caused to or, collapse of any part of the excavation and support work and in any case every seven days.

A written report giving full details of the examination including the name of the person who carried it out and the date must be made.

(iii)   All materials must be inspected before their use.

(iv)   The support work must be properly constructed and struts and braces securely fixed against accidental dislodgement.

(v)   If there is a risk of the excavation being flooded, ladders or other means of escape must be provided.

(vi)   If an existing structure is likely to be affected by the close proximity of an excavation, that structure must be shored or supported against collapse.

(vii) Excavations of more than 2.0 m in depth which are close to persons working at ground level or passing by, must be fenced off with guardrails.

When access to the excavation is required by plant or the movement of materials the guardrails may be temporarily removed but must be replaced as quickly as possible.

(viii) To avoid the collapse of the side of an excavation or materials falling on to persons working within the excavation, plant, materials and spoil must be kept clear of the edges.

The fundamental requirements of support work may be summarised as follows:

(a) To provide safe working conditions.
(b) To allow the efficient execution of both the excavation and permanent construction work.
(c) Be capable of being easily and safely removed after completion of the permanent work.

The term 'timbering' in the context of excavation work is a general term used to describe any form of support and may be timber, steel or concrete.

For small to average sized excavations and in particular, trench excavations, timber planks up to 75 mm thick and 900 to 1200 mm in length are used to support the sides of the excavation. Steel trench sheeting in lengths of 2.0 to 6.0 m are also used.

For large excavations steel trench sheeting or driven interlocking steel sheet piles are normally required. Reinforced concrete diaphragm walls can be used and have the advantage of forming part of the permanent work. Walings are normally timber or rolled steel sections. Horizontal strutting or inclined shoring are usually adjustable steel props.

There are basically three ways of supporting the vertical timbering to the sides of an excavation. They are:

1. Strutting.
2. Shoring.
3. Anchoring.

1. Strutting

Strutting is used in trench excavations and cofferdams. The forces exerted by the walls of the excavation are transferred to horizontal struts at right angles to the walls by way of the vertical sheeting and walings. The horizontal and vertical spacing of the struts will depend upon the force to be resisted and the strength of the struts, see Figure 29.

Structurally this method is very efficient but it does have the disadvantage that the struts may obstruct

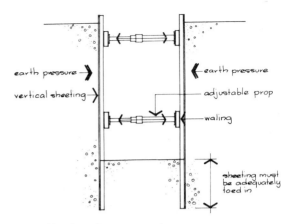

Figure 29 Strutting to excavations

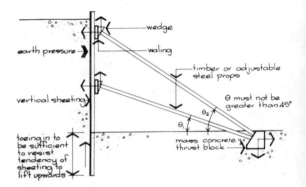

Figure 30 Shoring to excavators

the construction work and also limit the choice of excavation plant.

2. Shoring

On large excavations where room permits, the sides of the excavation may be supported by raking shores, see Figure 30. This method though is slowly being replaced by anchored systems of support and the diaphragm wall methods.

3. Anchoring

This method of supporting large excavations has the advantage of providing a working area free of the maze of struts and shores associated with the traditional methods. Steel sheet piling is driven to form a curtain around the area to be excavated. When the excavation work starts ground anchors are used to tie back the sheet piling.

Ground anchors consist of bars or strands and are stressed against an anchorage which holds the tie in the ground. A tensioning anchor is provided at the

24

other end. The ground anchor is positioned by boring a hole using a rotary percussion drill with casing in sand or gravel. An auger is used in clay soils. The bar or strands are then inserted and a predetermined length grouted up with neat cement, by pressure. When the anchorage has gained its strength the bar or strands are then tensioned to the required force. Figure 31 illustrates this method.

Another method of support which differs from the three previous types is that the wall of the excavation is supported by sheeting that spans horizontally between soldier piles, see Figure 32.

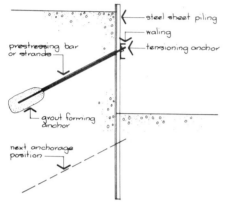

Figure 31 Anchored sheet piling to excavations

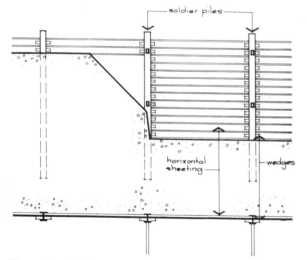

Figure 32 Soldier pile method of supporting sides of excavations. Elevation and plan

Soldier piles are usually H section steel piles driven at predetermined centres before excavation work is commenced. As the ground is excavated the horizontal sheeting (it may be timber, trench sheeting or precast planks) is lowered down and wedged tight between the soldiers.

A new method based on this system uses steel panels between the soldiers that lower themselves due to their self weight. The soldiers are normally strutted in trench excavations or anchored back in open excavations.

3.4 Backfill

The choice of backfill is an important aspect of earth works. The physical properties of the fill material and the method of backfilling used must both be considered. The physical properties are measured by the moisture content, compacted density and bearing capacity of the material. The moisture content of the fill will be laid down in the specification with the allowable variations, the actual figure during the work will depend upon the type of weather, the nature of the material, its grading i.e. the percentage distribution of the particle sizes, and the amount of compaction required.

The compacted density is measured by the Proctor compaction test (see 1.4) and the bearing capacity is measured by the California Bearing Ratio (see 1.4) test. These two tests can be carried out on site or in a laboratory, the site tests should only be used as control tests to compare with the laboratory results.

To ensure good compaction it is necessary that the fill is placed and compacted in thin layers. The contract specifications will sometimes state the thickness of each layer, the compactions to be achieved and the number of passes the compacting machine should have to make. If there is no such direction in the specification a thickness of layer for the compacted fill should be 300–375 mm for most soils. The advantages of carrying out the work in this manner

Figure 33 Compacting roller (Aveling Barford Ltd.)

are that the surface of the fill is constantly being renewed, hence avoiding rutting of the surface by the passage of the haulage vehicles. If watering of the fill is required the layers allow an even distribution of the moisture and by covering the compacted layer with new fill drying out by wind and sun is avoided.

Compacting is carried out either by hand-operated machines (on small areas) or towed or self propelled compactors such as vibrating rollers, sheep's foot rollers (Figure 33) and drum rollers of various kinds. Smooth surface rollers such as diesel rollers (steam rollers) are seldom effective for this type of work.

3.5 Backfilling procedures

The backfilling procedure for foundations and retaining walls should be carried out by the layer method (see 3.4 above).

3.6 Top soil

Top soil is the layer of soil which by its humus content (the fertile property made up mainly of rotted organic matter) supports vegetation. On average it forms the first 150 mm layer of soil in the UK.

If re-seeding of a site or part of a site such as embankments is required, the top soil which has no engineering value, must be replaced. As it forms a relatively thin layer it is a valuable material and must be carefully removed at the start of a contract to a special dump or spoil heap for later use. If it is not required for re-use the contractor may dispose of it by selling unless the contract specifies that it will remain the property of the client.

3.7 Imported backfill

In certain cases excavated material may not be suitable as a backfill and imported backfill must be used. In this event careful control of the quality and characteristics of the material must be carried out to ensure the fill complies with the specification. The mixing of two or more types of fill is not to be recommended as the proportion used and the mixing is difficult to control if done off site.

4. SHEET PILING

4.1 Definition

Sheet piling may be generally defined as closely set piles driven into the ground to keep earth or water out of an excavation.

4.2 Types of sheet piling

From the point of view of materials there are basically three types of sheet piling.

(i) Steel sheet piles.
(ii) Reinforced and pre-stressed concrete sheet piles.
(iii) Timber sheet piles.

(i) Steel sheet piles

Steel sheet piling is by far the most common form of sheet piling used in civil engineering work.

It is used for the construction of permanent retaining walls in docks, harbours and embankments, river bank and sea defence works. Steel sheet piles are also used to support the sides of trenches and open excavations and forming cofferdams.

A cofferdam is a temporary structure built for the purpose of excluding water or soil to permit construction to proceed without excessive pumping and, in the case of land cofferdams, to support the surrounding ground. A more detailed treatment of cofferdams will be found in the TEC unit Civil Engineering Technology IV (U78/444).

(ii) Reinforced concrete and prestressed concrete sheet piles

Generally used for permanent work that will be incorporated into the completed structure.

(iii) Timber sheet piles

Timber sheet piling is rarely used in the UK due to the cost, but it is used extensively in countries where timber is plentiful and steel or pre-cast concrete sheet piling is not readily available. In such areas timber sheet piling is mainly used for temporary structures in a similar fashion to steel sheet piling.

4.3 Construction methods

The construction and methods of driving sheet piling are described below.

(i) Steel sheet piles

The two main types of steel sheet piles used in the UK are 'Frodingham' and 'Larssen'. Although similar in application of use their difference is the position of interlock. When in position, both systems produce a wall which is trough-shaped in plan, see Figure 34. Fordingham piles are available in lengths of 9.0 m to 24.0 m and Larssen in lengths of 9.0 m to 26.0 m. In special cases the piles can be supplied in lengths of up to 30.0 m though handling such lengths of pile may prove difficult. An alternative to using such long lengths or in cases where the length of pile required has been underestimated. The pile may be increased in length by splicing on an extra length of pile with fishplates or by the use of site welding.

To keep the piles in position during driving it is essential to erect temporary guide walings. These can either be of trestle construction, or formed by the use of timber or steel piles, see Figure 35.

In general, the choice of timber or steel to form these temporary works will depend on relative cost and availability.

The piles should be supported or guided at two levels during driving by the walings, one pair being set as near to ground level as possible or in the case of marine works, just above the low water mark. To assist stability of the driving frame, the distance apart of the two sets of walings should be as large as is practicable. It is of the utmost importance that the walings be held rigidly in position. The vertical supports to the walings should be set at about 2.5 m to 4.0 m centres.

Sheet piling may be driven by any drop, steam, air or diesel hammer. The hammer is normally suspended from a derrick or crane and fitted to the piles by means of leg-grips or leg-guides to obtain correct positioning of the hammer on to the piles.

In certain circumstances, such as in urban areas, or where the work is close to an existing structure, the

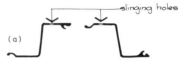

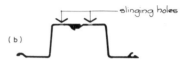

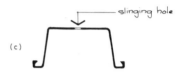

Figure 34 Steel sheet piling: (a) single piles; (b) Frodingham sections – double piles; (c) Larssen sections

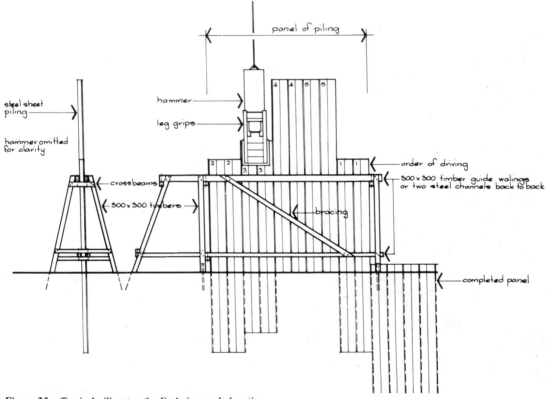

Figure 35 Typical piling trestle. End view and elevation

27

noise and vibration caused by the action of the hammer may be unacceptable. In such case an alternative method of driving is used.

The method most commonly adopted is to use hydraulic rams to push the piles into position. The Taywood Pilemaster, which is used throughout the UK and abroad, is a machine that works on this principle. Fitted with eight hydraulic rams the Pilemaster is capable of driving and extracting sheet piles with little or no measureable vibration and very low noise (69 dBA at a distance of 1.50 m from the machine), see Figure 36.

Figure 36 Taywood pilemaster (Terresearch Ltd.)

28

The most commonly used method of driving steel sheet piles is known as 'driving in panels'. This method is carried out in the following stages.

(a) A pair of interlocked piles are pitched and part driven to a depth level with the top walling and carefully checked for verticality.
(b) About six to twelve pairs of piles are then pitched and interlocked in position and held securely in the guide walings. Hence a panel of piles is formed, the first pair being partly driven.
(c) The hammer is now transferred to the last pair of undriven piles in the panel which are then part driven to about the same depth of the first pair. Again these must be checked for verticality.
(d) The remaining pairs of piles are then driven to their final level, starting with the pair adjacent to the last and working across to the pair adjacent to the first.
(e) The first pair is not fully driven as they will form the first guide pair of the next panel. The procedure is then repeated.

(ii) Reinforced and prestressed concrete sheet piles

The cross section of this type of pile is generally rectangular with either tongue and grooved or 'birdsmouth' shaped sides to form the interlock. The foot of the pile, as shown in Figure 37, is bevelled and feather edged to assist in the driving and also to help correct a tendancy for this type of pile to lean (or drift) as it is driven.

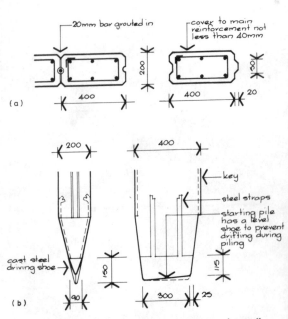

Figure 37 Typical precast reinforced concrete sheet pile details: (a) typical sections; (b) typical shoes

When the pile is to be driven in hard strata a metal shoe is used on the foot to protect it during driving.

As noted previously there is a tendancy for this type of pile to drift as it is driven; to counteract this problem the piles are driven one at a time working from left to right. The first pile must be plumb, of course, and careful control is required during the driving of it, the following piles will, as they are being driven, tend to drift towards the first pile hence forming a close fit.

If the completed work is to be watertight the joints are usually grouted up.

(iii) Timber sheet piles

The simplest form of timber sheet piling is rectangular planks mainly used as close boarding in trench excavations. Other shapes include birdsmouth (formed by bolting together double bevelled planks) and tongue and grooved, (formed by bolting together three rectangular planks).

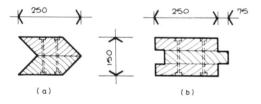

mild steel hoops placed around top of pile to prevent spread during piling; in soft ground conditions pile toe is bevelled and pointed, and tied with m.s. hoops as for top; in other ground conditions, cast steel point is attached to toe

Figure 38 Typical timber sheet piles: (a) birdmouth; (b) tongue and groove

Timber sheet piles are driven in a similar manner to R.C. and pre-stressed concrete sheet piles, see Figure 38.

4.4 Extracting sheet piling

Sheet piling is extracted by the use of a double-acting hammer in which the motion is reversed, or they can be jacked out. The Taywood Pilemaster extracts piles by reversing the hydraulic rams.

5. FUNCTIONS OF RETAINING WALLS AND PRINCIPLES INVOLVED

5.1 Factors affecting stability

The function of a retaining wall is to hold, in a permanent vertical position, a wall of earth that in an unsupported state would be unstable. This condition will occur where a significant and abrupt change in ground level is required. Figure 39 shows the position of forces acting on a retaining wall.

The stability of a retaining wall is dependant on three main factors.

(i) Overturning

A retaining wall will tend to overturn if the active pressure on the wall produces a stress at the toe of

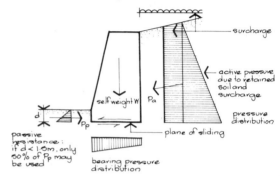

Figure 39 Forces acting on a retaining wall

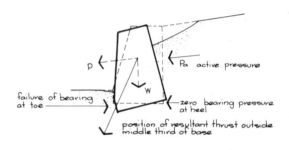

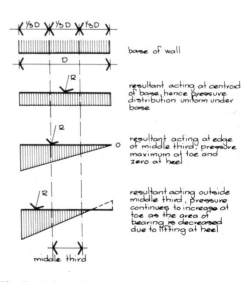

Figure 40 Retaining walls – overturning

the wall which the material supporting the toe cannot resist, or the resultant thrust of the forces acting on the base of the wall is outside the base length, see Figure 40.

It is a requirement of good practice to design the retaining wall so that there is a factor of safety of at least 1.5–2.0 against overturning.

(ii) Sliding

Retaining walls will fail by sliding if the frictional resistance between the underside of the base and the ground on which it sits is insufficient to resist the active pressure, see Figure 41.

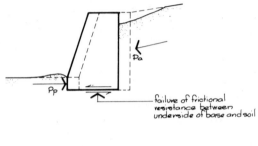

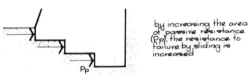

Figure 41   Retaining walls – sliding

The frictional resistance can be increased against sliding if the passive resistance of the soil in front of the wall is brought into the calculation.

The passive resistance available can be increased by introducing a downstand or rib on the underside of the base.

A factor of safety of 1.5 is normally required against this form of failure.

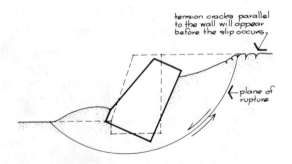

Figure 42   Retaining walls – circular slip

(iii) Circular slip

Retaining walls supporting cohesive soils such as clay, may fail by a plane of rupture forming along a curve that will carry the wall forward and tilting it back as it goes. This is caused by the shear failures of the clay, Figure 42.

5.2 Types of retaining wall

Retaining walls can be built using mass concrete, reinforced or pre-stressed concrete or steel sheet piling.

The choice of material used will, in most cases, be dictated by the form the wall takes. The various types of retaining wall may be classified by the following groups:

(i)   Gravity or mass.
(ii) Cantilever.
(iii) Counterfort.
(iv) Buttressed.
(v)   Anchored.

Groups (i), (ii), (iii) and (iv) are generally termed as 'free standing' and group (v) as 'tied-back'.

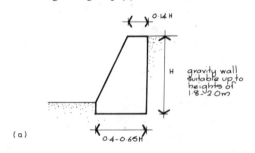

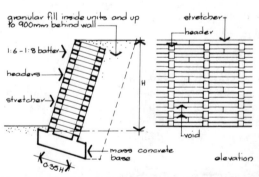

Figure 43   Gravity or mass retaining walls: (a) approximate proportions of a gravity releasing wall; (b) crib walling

(i) Gravity

Gravity walls rely on the mass of the wall to resist overturning and sliding. These are usually constructed in either mass concrete and, to a lesser extent brickwork. Figure 43 shows details of typical gravity retaining walls.

(ii) Cantilever

Cantilever walls rely on the strength of the stem at the base to resist the bending stresses induced by the pressure behind the wall. It must of course also resist overturning and sliding.

Reinforced concrete is the most commonly used form of construction. The position of the base in relation to the stem will depend upon the location of the wall, that is, it is not always possible to construct part of the base behind the wall although from a point of stability it is the position to be preferred. For details of a cantilever retaining wall see Figure 44.

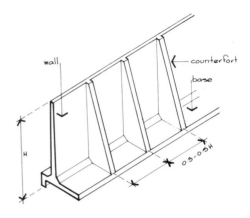

Figure 45   Counterfort retaining wall suitable for heights above 6−7 m

approximate guide to proportions

Figure 44   Typical R.C. cantilever retaining walls up to 5−6 m high

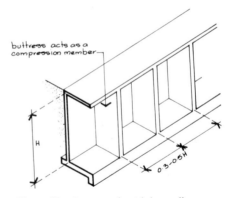

Figure 46   Buttressed retaining wall

(iii) Counterfort

Although similar in cross-section to the cantilever type. Counterfort walls span horizontally between the counterforts apart from the bottom 1.0 m of the wall which cantilevers from the base.

The counterforts act as cantilever T beams of tapering section and are usually spaced apart at distances of one half to one third the height of the wall. This type of wall is always constructed in reinforced concrete, see Figure 45.

(iv) Buttressed

For high retaining walls (above 6.0 m) where it is not possible to excavate behind the wall, a buttress can be used, see Figure 46.

This type of wall is cast against the face of the excavated soil. Like the counterfort type the wall slab spans horizontally.

(v) Anchored

Anchored retaining walls are generally constructed with steel sheet piling. The horizontal support is provided by means of anchorages near the top, as well as by penetration of the sheet piling into the ground.

The principal types of anchorage used are mass concrete block sheet piling balanced above and below the waling, and sheet pile cantilever. They may consist of isolated concrete blocks or pile groups or form a continuous wall, see Figure 47.

The walings are usually made up from two steel channels back to back and fixed on the inside face of the wall. As the tie rod is a most important part of the structure it should be protected from corrosion by wrapping in bitumen based hessian or grouted in a non-ferrous casing.

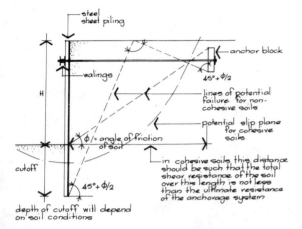

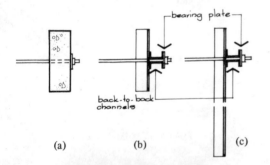

Figure 47 A typical steel sheet-piled anchored retaining wall showing approx. position of anchorage in cohesive and non-cohesive soils. Three types of anchorage: (a) mass concrete block; (b) balanced sheet piling; (c) cantilevered sheet piling

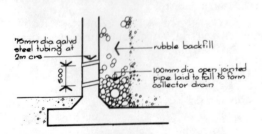

Figure 48 Draining details for retaining walls

5.3 Expansion and contraction joints

For continuous linear structures such as retaining walls expansion and contraction joints must be provided for reasons of durability. In brickwork vertical expansion joints should be provided at approximately 5.0–15.0 m centres.

Reinforced concrete walls in the UK require vertical expansion joints at 20–30 m centres and contraction joints at 5–10 m spacings. The reinforcement should not be continuous through the joints. For convenience, pouring of the concrete should be carried up to a contraction or expansion joint, to ensure good keying on commencement of pouring the concrete the next day.

If this is not possible a key should be formed by either casting a bevelled groove or by exposing the coarse aggregate by hosing with high pressure water.

5.4 Weepholes and drainage

To prevent a build up of hydrastatic pressure due to poor drainage behind a retaining wall, weepholes for drainage must be provided for. Figure 48 shows a typical detail.

6. UNDERPINNING AND DEWATERING

6.1 Underpinning

The purpose of underpinning is to transfer the load carried on a foundation from its existing bearing level to a new level at a lower depth. This may be necessary for one, or a combination of, the following reasons

(i) Settlement of foundations.
(ii) To increase the load bearing capacity of the foundations.
(iii) To allow works to be carried out below or adjacent to the foundations.

6.2 and 6.3 Underpinning operations

Before underpinning work is begun the ground conditions must be investigated to identify the conditions responsible for any settlement. This will enable an appropriate system of underpinning to be chosen, also a general picture of the ground conditions and a measure of the bearing capacity of the soil on which the underpinning is to be supported.

It is very important that a full survey is prepared of the structure to be underpinned and any adjacent structures that may be affected by the work.

Records should be made of the levels of the floors and the inclination of the walls, marking, noting and photographing any cracks and/or defect of the structure. Glass 'telltales' (a 'telltale' is a strip of glass that is placed across a crack) and datum points should be placed where necessary for observing the movement of cracks and settlements. The report of the survey must be agreed by the owners of the properties, the contractor and the engineers as a true record of the condition of the structure before work is commenced.

During the work a constant check must be kept on the structure by checking the original datum points and 'telltales'. It is also advisable; where possible, to remove as much 'live load' from the structure to minimise the load to be supported during underpinning operations.

Two methods of underpinning strip foundations supporting a wall are described below.

(i) Traditional

The simplest form of underpinning to lower the level of an existing strip foundation is to carry out the work in a series of legs or pits. The length of each leg will depend on the spanning ability of the existing foundation. Generally for brickwork walls of normal type it is recommended that each leg should be about 1 m to 1.4 m in length but for walls capable of 'arching', a greater leg length may be used, see Figure 49.

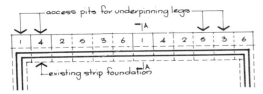

the maximum sum of unsupported lengths should not be greater than one quarter of the total length of the structure or one sixth of the total length if the wall shows signs of weakness or is heavily loaded

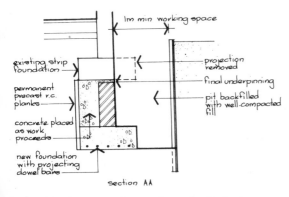

Section AA

Figure 49 Sequence of excavating underpinning legs

Once the leg length has been excavated the underside of the wall or foundation should be cleaned and levelled ready for pinning. The formation level of the leg should not be exposed until work is to commence, this requires that the last 100 mm or so of excavation be left until the levelling etc., of the old foundations has been completed. The leg should be constructed as quickly as possible up to within 75–150 mm of the underside of the old foundation ready for final pinning.

When the leg has reached a sufficient strength to support the load to be placed upon it the final underpinning can commence. This is carried out by ramming in a fairly dry concrete until hard up against the underside of the old foundation. The sides of the leg should be keyed to bond in with the adjacent legs and dowel bars placed in the ends of the new strip foundation.

The new series of legs should not be started until the preceding underpinning is completed and finally pinned.

(ii) Piled underpinning

If the depth of the formation level of the new foundation is too deep to use the above method, a piling arrangement may prove more economic.

These methods may involve some vibration, which would prove to be undesirable in many cases of underpinning. It is therefore obvious that the condition of the structure and foundation must be taken into consideration before a decision can be made on the method to be adopted.

A system has been developed in which vibration is virtually eliminated. The method makes use of friction piles of various diameters. A hole is drilled at an angle, either directly through the concrete foundation or through the brickwork directly above to a predetermined depth. The pile is completed by the introduction of a cage of reinforcement in the hole and the injection of cement sand grout pumped in under pressure.

Specialist contractors are continually improving their techniques in this area. Readers who require a more in depth knowledge of this objective are advised to obtain further information direct from these specialists.

6.4 Reasons for dewatering subsoil

In excavation work the ingress of surface and ground water can cause serious problems such as the undermining of supports to the sides of the excavation and the very costly delay of the work being held up due to flooding. There is also the general problem of working in waterlogged conditions; it is therefore important to control, or exclude, the flow of surface and groundwater in the excavation.

The methods used to control groundwater are generally termed as geotechnical processes, of which there are many. Such methods include the lowering of the groundwater, changing the physical characteristics of the subsoil, freezing the groundwater and the use of compressed air.

Before a final choice can be made on the techniques to be used, a good knowledge of the ground and water conditions of the site, and possibly surrounding area, will have to be obtained. This must be carried out before work is commenced as unnecessary delays will be caused if the problem of groundwater is only considered when difficulties arise during the construction work.

6.5 Information required

(i) Preliminary site investigation

As previously described in Topic Area A, a site investigation is carried out to obtain a general picture of the underlying subsoil strata and water level of the site. The investigation will establish the likelihood of the need for dewatering operations; if this is seen to be so then a more detailed investigation will be required.

(ii) Detailed investigation

This investigation is required to allow a detailed study of the soil profiles groundwater conditions and, if relevant, a general history of other excavation and foundation works that have been carried out in the area. This latter point can be of great use in assessing the problems to be faced.

The types of soil that make up the strata, underlying strata and their particle size distribution will have an important bearing on the choice of dewatering method to be used. When fine grain soils are encountered it is recommended to obtain continuous samples.

Particle size analysis should be made on representative samples from all silts, sand and gravel layers. In the case of silts and fine sands the samples should be undisturbed.

If dewatering is to be carried out in the vicinity of existing structures the compressibility of the various strata must be determined in order to ascertain whether excessive settlement is likely to occur. The settlement is caused by the reduction of pore water pressure in a compressible strata. If this were to happen serious damage may be caused to the existing structures due to the settlement of their foundations.

A study of the groundwater flow pattern is of great value when considering the layout of the groundwater installations. The purpose of the study is to establish the source of the groundwater and how it flows in the region of the site. Groundwater is normally found in pervious strata due to the percolation of rainfall and run-offs and near streams and rivers.

For civil engineering purposes there are two types of groundwater problems:

(a) Confined.
(b) Unconfined.

(a) Confined

Groundwater is said to be confined when the layers in which it lies has along its upper flow boundary a layer of low permeability. If the source of the groundwater is above the level of the area of investigation the water head in the previous layer at this point may be above the ground level. This condition is known as artesian, see Figure 50.

(b) Unconfined

When the upper flow boundary of the groundwater is not confined by a layer of low permeability it is said to be unconfined. The groundwater pressure at this boundary will be equal to atmospheric pressure, see Figure 51.

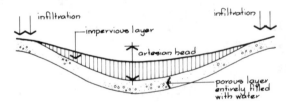

Figure 50 Confined flow

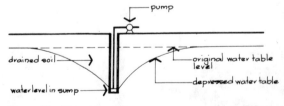

Figure 51 Unconfined flow

In the case of confined groundwater it is necessary to measure the water pressure in the previous strata at suitable intervals across the site, in order that the maximum water pressures be ascertained. It is also

34

required to build up a picture of the ground water flow across the site; to act as a control to measure changes in water pressure due to pumping test, and groundwater lowering operations. Also to ensure that the groundwater pressure does not exceed values which could cause damage to temporary or permanent works.

Groundwater pressure is measured by devices known as piezometers. The apparatus basically consists of a ceramic or porous plastic piezometer tip which is connected to a p.v.c. or steel tube standpipe. The water pressure is measured by either connecting the standpipe to a water level indicator or a monometer gauge, a device for measuring water pressure difference.

If the apparatus is to remain *in situ* during the construction work and a borehole was used to position it, the area of the tip should be filled with sand and sealed with a layer of Bentonite. The borehole is then filled in or grouted up, see Figure 52. It should be noted that the choice of piezometer tip requires specialist knowledge which is beyond the scope of this topic area.

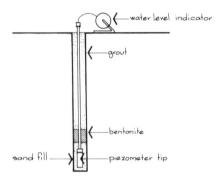

Figure 52 Diagrammatic view of piezometer *in situ* (Casagrande method)

It is also necessary to measure the rate of flow and permeability of the groundwater. The rate of flow is controlled by the permeability of the soil and is expressed by the equation (Darcy's Law):

where $$v = ki$$

v = the superficial velocity of flow through the soil;
i = the hydraulic gradient;
k = the permeability.
k is measured in m/s and depends chiefly on particle size and grading. Typical values of permeability for soils range from 1×10^{-5} m/s for coarse sand to 1×10^{-9} m/s for a clay.

Permeability can be measured either *in situ* or in a laboratory. The usual site tests are full scale pumping tests, rising and falling head tests in a bore hole, and the calculation of flow conditions measured in boreholes.

Having obtained the above information the final factors to be considered before a decision on the method to be adopted will include the size and shape of the excavation, the length of time it will be open and the overall economics of any particular choice.

6.6 Pumping from sumps

A sump is a pit in which water collects before being pumped out.

Traditionally sumps are sited within the area of excavation, but when large quantities of water have to be pumped it may be more convenient to form the sump just outside the excavation area. This will also avoid the risk of damage to supporting timbers in the excavation by erosion of the soil at formation level.

Except on very small sites at least two sumps should be used. The sump should be excavated to the full depth required to drain the excavation before the main excavation reaches the level of groundwater, and maintains in its original form until completion of the construction work.

To prevent loss of soil due to pumping, a filter medium can be placed in the sump. In order to do this a cage of perforated metal is placed in the bottom of the sump and the space between the cage and the ground filled with a graded gravel filter material. As the gravel is placed the timber or steel supports to the sides of the supports are withdrawn.

By excavating the sump, before the main excavation below the groundwater level is commenced, the groundwater can be kept below excavation level at all stages and any difficulties that might arise due to the soil or groundwater will be identified during the construction of the sump. This will allow the engineers and contractor to make any changes in the excavation scheme before the main work begins.

Drainage channels should be dug to falls leading to the sumps to prevent water standing on the surface of the excavation. The falls must not be too slow to allow silting up or too steep to cause erosion. The most efficient method is to use open-jointed earthenware pipes in the channels backfilled with a filter material.

If the excavation is to be taken below a pervious layer into an impervious one it is good practice to have the sumps at the base of the pervious layer, being fed by channels stepped back into the perimeter of the excavations. This is known as a 'garland drain'.

The advantage of doing this is that the formation level of the main excavation will be kept free of water

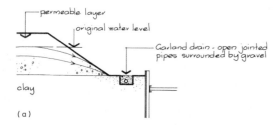

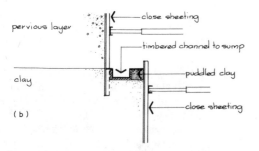

Figure 53 Types of Garland drain: (a) at top of excavation with sloping face; (b) in fully supported excavation

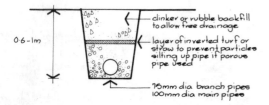

Figure 54 Typical section through land drain trench

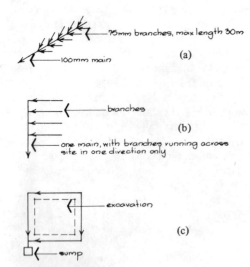

Figure 55 Typical layouts of land drains:
(a) herringbone; (b) grid iron; (c) moat or cut-off

which could lead to it being cut up by the passage of plant, and a reduction in the size of pumps required, see Figure 53.

6.7 Land drains

Land drains are used on construction sites to control surface water over the site and in certain cases to prevent surface water run-off entering an excavation.

They are normally open joined porous clay, concrete or perforated pitch fibre pipes. Vitrified or polythene pipes are also used. The choice of pipe is mainly based on the availability of any particular type. The method of laying the pipe in a trench is extremely important. Figure 54 shows a cross-section through a typical land drain trench.

If porous pipes are used there is the risk of the pipes being silted up but with correct backfilling the risk is small. Figure 55 shows some typical layouts of land drain systems. The whole system will generally discharge into either a soakaway or a watercourse via a catchpit.

7. PILING TYPES AND INSTALLATION OF BEARING PILES

7.1 Definition

A piled foundation may be defined as a means of transferring the load from a structural member (such

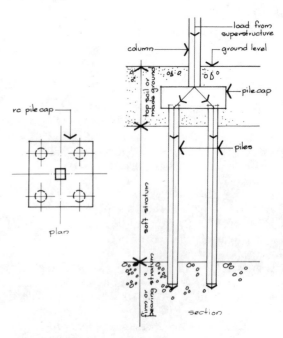

Figure 56 Typical four pile group

as a column or wall) to a firm stratum at some depth below the base of the structure.

Apart from very simple situations, or where large-diameter piles are used, it is usual to employ piles in groups. Figure 56 shows a sketch of a typical four pile group. It can be seen that the load from the column is transferred through the reinforced cast *in situ* concrete pile cap in such a way that each pile receives an equal share of the load. The pile in turn acts as a column and transmits the load to the surrounding strata.

A pile may transmit their loads to the surrounding strata by:

(i) End bearing

In this case the piles transfer their loads from the superstructure through soft ground, and are supported by a firm stratum at a lower depth. It is usual for the end of the pile to enter the firm stratum for a short distance.

(ii) Friction

Where a firm strata cannot be reached at a reasonable depth the piles rely for support on the side resistance of the soil over a considerable proportion of their length.

It should be noted that in practice, piles are supported by a combination of (i) and (ii). The two types of pile are illustrated in Figure 57.

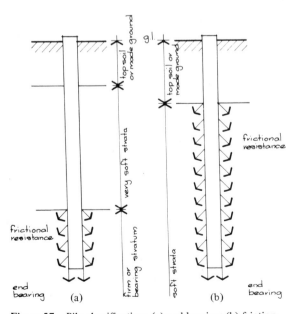

Figure 57   Pile classification: (a) end bearing; (b) friction

7.2 Piled foundation

The choice of a foundation for any particular structure will be influenced by many factors. Some of these will be economic influences, but the main factor will be the type of sub-soil encountered.

In Topic Area A reference was made to the importance of a soil report to determine the bearing capacity of the soil. When this report is received by the civil or structural engineer a decision is made as to the type of foundation to be used. However it should be noted that the report itself will suggest a certain type of foundation for the structure, but it is the engineer who is responsible for the project who will make the final decision.

If a four-storey structural frame is considered, in either reinforced concrete or structural steel, and the soil report indicates that a firm stratum will not be reached until 5.0 m or more below ground level, then a piled foundation would probably prove to be the most economic. In certain circumstances such as a high water table, or presence of a highly compressible soil near the surface, overlying a firm stratum, then a piled foundation might be considered at a shallower depth.

As stated previously, either end bearing or friction piles could be used in certain circumstances, but in this case end bearing piles only are considered.

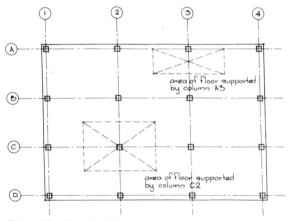

Figure 58   Framing plan

Figure 58 shows a framing plan for the structure considered and indicates the position of the columns.

If piles of the same type and diameter are used throughout the structure, the pile groups supporting the columns on grids (B2) (B3) (C2) (C3) would have the highest number of piles, as these columns carry the greatest load. The pile groups on grids (A) and (D) would not require the same number of piles as the column loads would be less.

The arrangement of various pile groups are shown in Figure 59. The spacing of the piles are dependent on the soil conditions encountered, although there must be a minimum distance between adjacent piles otherwise they will tend to act against each other. The approximate minimum spacings for end bearing piles are 2.5 × the distance of pile, so that if 400 mm diameter piles are used the minimum spacing would be 2.5 × 400 = 1000 mm.

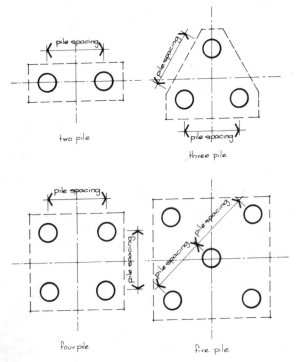

Figure 59   Various pile groups

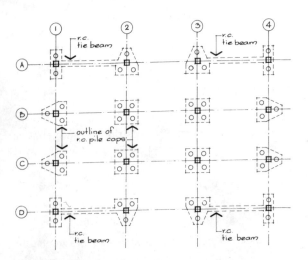

Figure 60   Piling layout

Figure 60 shows a possible pile layout for the proposed structure. It should be noted that the two pile groups are unstable about an axis, and should be tied back to a stable pile group by the use of a substantial reinforced concrete beam. In practice it is usual to provide the beams between all the pile groups. Their purpose is to provide support for the external cladding and in some cases the ground floor slab, if it is designed as suspended.

7.3 Types of pile
Piles can be divided into two main types:

(i)   Displacement piles, and
(ii) Replacement piles.

(i)   Displacement piles

In this case the pile is driven into the ground displacing the soil but not removing the material. The disadvantage of this type is that driving the pile could cause unacceptable noise and in some cases excessive vibration which could cause damage if driving takes places close to an existing building. The displacement pile can be sub-divided into three types — performed, partially preformed and driven cast *in situ*. These types are discussed in later paragraphs.

(ii)   Replacement piles

With this type, a hole is formed in the ground by removing the soil and replacing with *in situ* concrete. The advantage of this method is that as the boring operations are only mildly percussive, piling can take place close to an existing building, or in some cases, where headroom is limited.

The construction and installation of piles is usually carried out by a specialist contractor who usually employs his own particular type and method to a given contract. Three different types are only discussed in this book, but suggested further reading is given at the end of this topic area.

(iii) Prefabricated driven piles

This preformed displacement type of pile is used where ground conditions are suitable, for example where soft ground overlies a firm strata. The piles can be manufactured from different materials such as timber, steel or concrete. Only the concrete type of pile is considered here.

Concrete piles are usually square in cross section and manufactured in a casting yard, which in the case of a very large project could be on the site, thereby reducing transporting difficulties.

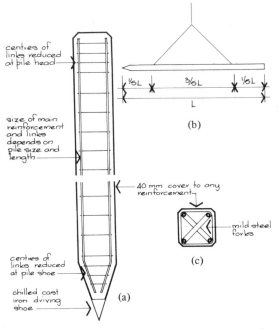

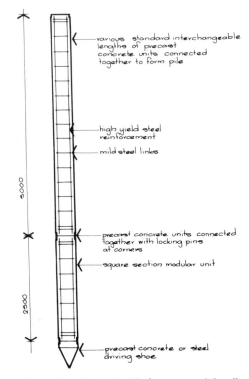

The disadvantage of this type of pile is the waste involved in cutting off excessive lengths of pile, or by lengthening piles found to be too short. As previously mentioned, excessive noise and vibration could be a problem.

This type of pile has been largely superseded by the partially preformed type developed by some specialist piling contractors. Typical example of this type is the West's 'Hardrive' pile shown in Figure 63. This type of pile has largely been developed to avoid some of the disadvantages of the preformed type.

Figure 61 Precast reinforced concrete pile: (a) elevation; (b) lifting position for precast pile; (c) section

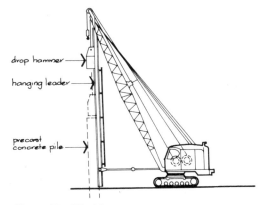

Figure 62 Pile driver

Figure 63 West's 'Hardrive' precast modular piles

Figure 61 shows details of a typical pile, the concrete used in the manufacture, is usually of high strength to resist the stresses set up by the heavy driving involved. As the piles are cast in a horizontal position, reinforcement must be positioned so as to resist the bending stresses when the piles are lifted. In addition, links must be provided throughout the length of the pile, the centres of which must be reduced at the head and shoe to resist the effects of the driving. A helmet must be provided over the head of the pile to protect it during the driving operations.

The installation of these piles is carried out by using a pile driving rig and a suitable hammer. The weight of the hammer will depend on the size of pile to be driven, see Figure 62.

(iv) Driven in situ concrete piles

This displacement type of pile is suitable where a weak strata overlies a firm stratum, or where a high water table is indicated.

The method employed is that a steel tube with a closed end is filled with a suitable semi-dry concrete mix (which is commonly known as a 'plug') to a depth of 1.0 m. The tube is then driven into the ground by a drop hammer operating against the 'plug', when a suitable depth is reached, any surplus tubing is cut off. A cylindrical cage of reinforcement is lowered into the tube, as the concrete is placed the tube can either be removed or left in position depending on the method employed. The difficulty is to ensure that

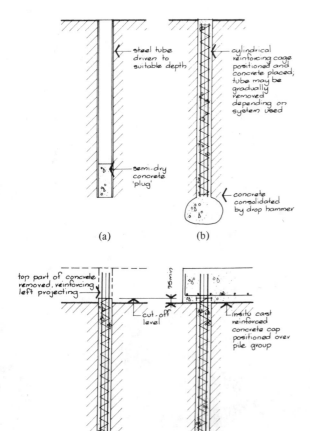

(a)
- steel tube driven to suitable depth
- semi-dry concrete 'plug'

(b)
- cylindrical reinforcing cage positioned and concrete placed, tube may be gradually removed depending on system used
- concrete consolidated by drop hammer

(c)
- top part of concrete removed, reinforcing left projecting
- cut-off level

(d)
- in situ cast reinforced concrete cap positioned over pile group
- 75min

Figure 64 Sequence of installation of driven *in situ* concrete pile

Figure 65 Constructing piles in a restricted area with limited headroom (Frankipile Ltd.)

adequate cover to the reinforcement is maintained. The disadvantage of this type of pile is that the plant used requires considerable headroom. Figure 64 shows the installation stages of a driven *in situ* concrete pile and its connection to the pile cap.

(v) Percussion bored piles

The advantage of this type of replacement pile, is that it is very useful on sites where vibration must be kept to a minimum because of the proximity of buildings, tunnels, sewers, etc. Percussion bored piles can be used virtually in any situation, see Figure 65, as the plant required is light and easily portable.

The method of construction is that a hole is formed by dropping a digging tool, which is usually a percussion cutter, when the tool is lifted it brings some of the soil with it. As the hole is formed steel tubes made up in 1.0 m lengths and screwed together are

driven into the hole. When a firm stratum has been reached a cylindrical cage of reinforcement is lowered into the hole, as the concrete is placed the steel tubes are gradually withdrawn. Once again the problem is to ensure that adequate cover is maintained to the reinforcement. This method is only suitable when the number of piles employed is small as the rate of boring is very slow.

(vi) Rotary bored piles

This type of pile is generally used where there is a clay subsoil. The soil is extracted by the use of an open flight helical auger, which when rotated fills the flights with spoil. When the flights are filled the auger is removed from the hole and the soil deposited, the operation is then repeated until a suitable depth is reached. Steel tubes are used if soil conditions demand to prevent the ingress of soil during boring. The tubes

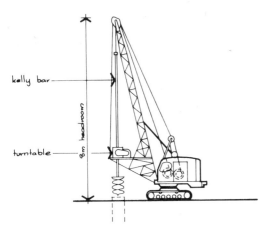

kelly bar

turntable

8 m headroom

Figure 66 Pile drill

are removed after the reinforcing cage has been positioned and as the concrete is being poured.

As the rotary bored pile involves the use of heavy equipment it is obvious that they are not economic for small sites, and as Figure 66 shows, the equipment which can reach a height of 8.0 m or more, is only suitable for open sites.

The rotary bored method is suitable for small diameter piles employed in groups or for the single large diameter pile. The economic and engineering considerations involved in choosing between small or large diameter bored piles are beyond the scope of this book.

Further reading
Readers who require a more in depth knowledge of some of the objectives in Topic Area B may find the following reading list useful.

B.3 Stubbs, Frank W. Jr., *Handbook of heavy construction*, McGraw Hill Publishing Co.
Paxton, J. M., *Manual of civil engineering plant and equipment*, Applied Science Publishers.
Foster, Jack Stroud and Harrington, Raymond, *Mitchell's building construction structure and fabric: Part 2*, B. T. Batsford Ltd (1976).
B.4 *B.S.P. Pocket Book*, British Steel Piling Ltd.
B.5 Tomlinson, M. J., Foundation design and construction, Pitman (1969).
B.6 Foster, Jack Stroud and Harrington, Raymond, *Mitchell's building construction structure and fabric: Part 2*, B. T. Batsford Ltd (1976).
CP 2004: 1972. Foundations, British Standards Institution.
B.7 West, A. S., *Piling practice*, Butterworths (1972).
Faber, John and Johnson, Brian, *Foundation design simply explained*, Oxford University Press (1976).

Topic Area C – Superstructure

8. SCAFFOLDING – TYPES, EXECUTION AND DISMANTLING

8.1 Types of scaffold

A scaffold is a temporary structure which can be of tubular steel, tubular aluminium alloy or timber. However, it should be noted that a scaffold of timber, other than the working platform itself, is very rarely used in this country, although it is still used extensively abroad.

The tubular steel or aluminium members are usually 50 mm outside diameter, but the steel members are much stronger, and therefore used for scaffolds which will support heavy loads. The aluminium members are used for a light scaffold which may be required for maintenance work.

The advantage of aluminium members is that they are virtually maintenance free, whereas the steel members will require some protection, to guard against corrosion.

The timber working platform usually consists of softwood boards 225 × 38 mm in cross section and up to 4800 mm in length. The width of a scaffold varies from four to six boards depending on its use.

The scaffold terms and fittings are referred to in the TEC unit Construction Technology II (U75/074).

There are two forms of scaffolding.

(i) Putlog scaffold.
(ii) Independent scaffold.

(i) Putlog scaffold

This form of scaffold consists of a single row of standards, erected at a distance from the structure, depending on the width of the working platform required. The standards are joined together with ledgers and tied to the structure with cross members called putlogs. This type of scaffold is erected as the structure rises, see Figure 67.

(ii) Independent scaffold

Figure 68 shows an independent scaffold which has two rows of standards tied together by cross members called transomes, and joined together with ledgers. As the scaffold is independent, it should be securely tied to the structure at frequent intervals, usually across a window opening. If a suitable opening is not available,

the scaffold should be strutted from the outside. This type of scaffold is used when work is to be carried out to an existing structure.

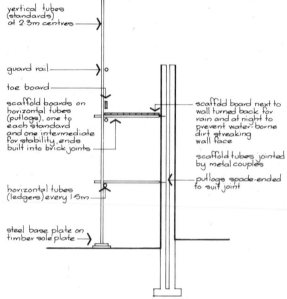

Figure 67 Putlog scaffold

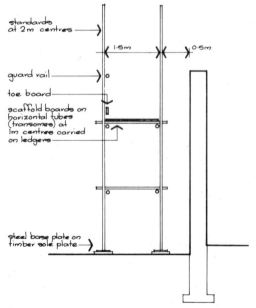

Figure 68 Independent scaffold

A mobile independent scaffold can be made from aluminium alloy tubes which is constructed in the form of a 'tower'. Special attention should be made to the bracing of the scaffold. The standards can rest on base plates or wheels. This form of scaffold can be used for internal work, or as a support for formwork.

The erection and dismantling of scaffolds and ladders is covered by the *Construction (Working Places) Regulations 1966.* Particular attention should be given to Regulation 22 which covers the inspection of scaffolds, and Regulation 39 which refers to the information which is required from such an inspection.

9. READY-MIXED CONCRETE – QUALITY CONTROL, TRANSPORTATION AND CURING

9.1 Tests for concrete

However carefully the batching and mixing for the site mixed or ready mixed concrete process may have been carried out, it is usually necessary to test concrete, to determine whether the desired quality is being uniformly maintained.

To assess the workability of fresh concrete BS 1881: 1970 recommends three methods. These are the slump, compacting factor, and the Vebe Consistometer tests. The last one cannot be considered as a site test, except in the case of a very large site, where the necessary apparatus may be available.

Although the tests are described in the TEC unit Science and Materials II (U75/042), the first two are also briefly described below.

(i) Slump test

The specimen is formed in a circular metal mould as shown in Figure 69. The mould is filled with concrete, and compacted with a tamping rod, as described in BS 1881: 1970. When the mould is filled with concrete, it is raised and placed alongside the specimen. The slump is measured by placing the rod across the top of the mould and measuring the distance the concrete has settled, see Figure 70.

(ii) Compacting factor test

The apparatus for this test is shown in Figure 71. The upper hopper is filled with fresh concrete, and allowed to fall through the hinged flap in the base, a standard drop into the lower hopper. The specimen concrete is allowed to fall into the cylinder by opening the hinged flap in the lower hopper. The cylinder is then weighed and the weight of concrete found by subtraction. This is then compared with the weight of the materials necessary to fill the cylinder without

Figure 69 Slump cone

Figure 70 Measuring the slump
(Cement & Concrete Association)

voids. Thus a ratio or factor is obtained which indicates workability, increasing as the factor approaches unity.

The slump test provides a means of controlling the water content of successive batches of the same mix, providing there is no substantial change in the aggregate gradings. The compacting factor test is more useful for mixes of low workability.

Concrete resists compression better than any other type of stress. In many cases, therefore, the compressive strength of concrete may be its most

Figure 71 Compacting factor apparatus (Cement & Concrete Association)

Figure 72 Steel mould (above) and concrete cube (below)

important property. In order to assess the strength of hardened concrete BS 1881: 1970 describes the following test.

(iii) Cube compressive test

The concrete samples are taken from the place of mixing, or in the case of ready mixed concrete, when the delivery is made to the site. The steel mould size is normally a cube of 150 mm (see Figure 72) but in certain circumstances this can be reduced to 100 mm. BS 1881: 1970 describes the method of compacting the concrete in the mould, and how the cubes are stored in a tank of water.

The cubes are tested wet, and directly between the platens of the testing machine. It is usual to test cubes at certain preferred ages. CP 110: 1972 classifies portland cement concrete on the basis of a twenty-eight day strength.

9.2 Storage and mixing

For normal site conditions, the materials to be used for concrete production, are delivered to the site ready for batching and mixing. It is essential that the materials are not allowed to deteriorate after being delivered to the site.

Cement is usually delivered to the site in 50 kg bags. This is a convenient size for handling and storage, although increasing use is being made of bulk handling equipment. The advantage to be gained from this method of storage is dependent on the size of the site and the amount of concrete to be produced.

When cement is delivered to site in bags it should not be stored directly on the ground as it would soon become damp. The bags should be placed on a raised platform and covered with waterproof sheets. The deliveries should be so arranged that the cement is not subjected to prolonged storage.

Aggregates are delivered to site in at least two sizes — classified as course and fine. It may be necessary for the course aggregate to be delivered in a larger number of different sizes. These could be 40 mm, 20 mm and 10 mm depending on the use required on site. The aggregates should be kept separated on the site, by providing division walls, each size being clearly labelled. The ground on which the aggregates are to be stored should be given a layer of lean concrete, laid to falls. This is to ensure that the aggregates do not become contaminated with mud.

9.3 Methods of transportation

The two primary considerations in selecting concrete placing equipment are:

(i) Concrete specification;
(ii) Site conditions.

The specification would include such factors as strength of concrete, type and size of aggregate and workability.

The site condition factors would include such things as: methods of batching, size and type of mixers, type of formwork and transporting distance.

It is therefore essential that to prevent, or at least minimise, segregation and prevent the premature compaction, the freshly mixed concrete should be transported as quickly as possible to the point of placing. The equipment must allow a clean transfer of the concrete mix. There are many methods of transporting fresh concrete, but only three are considered in this book.

(i) Dumpers

The powered dumper is a variation on the wheelbarrow theme, see Figure 73. Since they have two pairs of wheels and a lower centre of gravity, the dumper is more stable. It can carry more material than the wheelbarrow and obviously the transportation and placement of the concrete is quicker.

Figure 73   Powered dumper (Thwaites Ltd.)

(ii) Skips and buckets

When a crane is used for moving concrete, a suitable skip or bucket is required. The mix can be delivered directly to the point of placing in a single operation, without intermediate handling, see Figure 74. Large and small quantities can be handled to almost any height. This method is therefore the most efficient means of handling concrete for most structural work.

(iii) Agitators

Depending on site conditions, it may be advantageous to use ready mixed concrete. One method used for delivery is by a mobile agitator, see Figure 75. The agitator carries a batch of fresh concrete, which has been completely mixed at the depot and is kept on the move in a drum which is rotating slowly during

Figure 7 4   A crane-lowered skip

45

Figure 75   Concrete truck mixer (Winget Ltd.)

the journey. The maximum journey time for this method is 30 min or a distance of up to 10 miles from the depot to the site.

9.4 Compaction

Concrete on small contracts is often compacted by hand. Concrete mixes with a medium to high workability can be satisfactory compacted by tamping or puning the concrete with a rod or similar tool. To avoid leaving uncompacted regions below the surface, the concrete should be compacted in layers of not more than 300 mm for mass concrete and 150 mm for reinforced concrete.

A vibrator is a mechanical oscillating device of high frequency that is used to consolidate the fresh concrete. The use of mechanical vibration allows leaner mixes to be used to produce concrete of a given strength.

An internal poker vibrator is considered to be the best type. It will consolidate concrete very well if properly used. It should never be left vibrating in one place, any longer than 5–10 sec, as the concrete will tend to segregate. This type of vibrator must never be used to move concrete.

An external clamp-on type of vibrator is fixed to the formwork, and the energy of vibration must be transmitted to the concrete through the forms. It must be remembered that a stronger and more rigid formwork must be provided particularly when an external vibrator is to be used.

9.5 Curing

After concrete has been placed and compacted, it must be allowed to mature, and harden under satisfactory conditions of moisture and temperature, before its manufacture can be regarded as complete. This is because its hardening progress depends on the hydration of the cement and this can only proceed if the temperature is within certain limits, and if water is freely available to combine with the cement.

Portland cements hydrate more quickly at higher temperatures. Therefore appropriate steps must be taken to maintain a reasonable temperature when concreting in cold weather.

In practice the first essential of good curing is to ensure that the concrete remains wet long enough to allow its strength to develop sufficiently at early ages. If it dries out rapidly, it shrinks, and tensile stresses are set up if the shrinkage is restrained. Shrinkage restraint can occur, if the surface dries out rapidly, leaving the interior of the concrete in a wetter state than the outside. This frequently happens in practice and produces surface cracking. This does not usually represent structural failure, but the cracks can spoil the surface appearance of the concrete.

9.6 Methods of curing

Specifications for concrete construction should require provision for adequate wet curing CP 110: 1972 contains a clause relating to such curing.

There are many methods of curing but only the wet method is discussed in this book.

(i) Sprayed water

Water is applied by sprinkling the concrete, or by allowing a slow stream of water to run over it. Both of these methods are good if properly controlled. In both cases the entire surface of the concrete should be kept constantly wet. Care should be taken that the application of the water does not wash away the subgrade, thereby undermining the concrete. Water application should be maintained for at least 72 hours. This is difficult to achieve, so these methods are often combined with other methods.

Wet moisture-retaining fabric coverings should be placed as soon as the concrete has hardened sufficiently to prevent surface damage.

(ii) Straw blankets

Straw blankets can be used to cure flat surfaces. It should be placed in a layer at least 150 mm thick and kept constantly wet. A major disadvantage is the possibility of discolouring the concrete. This effect should be considered when using this method of curing.

(iii) Polythene sheeting

This type of sheeting is light and completely impervious to moisture. It is flexible and reusable, and the weight is constant whether wet or dry. Polythene sheeting requires no cutting to fit around corners or odd shapes. The use of this type of sheeting should be avoided on coloured concrete, unless the polythene can be kept from touching the surface of the concrete.

The above methods generally apply to ground or suspended slabs. The method of curing structural concrete members, such as beams, columns, or walls etc. is that once the formwork has been removed, the exposed concrete is covered with damp hessian sheets.

10. CONSTRUCTION OF STEEL AND IN-SITU REINFORCED CONCRETE FRAMES FOR STRUCTURES UP TO FOUR STOREYS HIGH

10.1 Structural steel sections

Structural steel sections are used in the construction of all types of structures from the simple single storey shed type, to multi-storey structures.

The sections may be classified as follows: universal beams; universal columns; joists; channels; equal and unequal angles; rolled and structural tees; circular and square hollow sections and castellated universal beams. Some of these are shown in Figure 76.

The shape and dimensions for the beam, column, joist and channel sections are specified in BS 4 Part 1: 1972 while the metric angles and hollow sections are described in BS 4848: Part 4: 1972 and Part 2: 1975 respectively.

The properties and safe loads for the various sections are tabulated in the *Structural Steelwork Handbook* (1978) published by the British Constructional Steelwork Association Ltd. The book is arranged in four parts, preceded by explanatory notes: Part 1 deals with dimensions and properties, including some tables of general information. Parts 2, 3 and 4 contain safe load tables for simple structural members in steel grades 50, 43 and 55 respectively, the pages for each grade being distinctively coloured.

It can be shown that the most economic shape for a steel beam is an I section, which has the maximum strength at the top and bottom of the section to resist bending, and the minimum strength at the neutral axis where the bending stress is zero.

The universal beams listed in the Handbook vary from a maximum size of 914 mm × 419 mm to a minimum size of 203 mm × 133 mm. A limited range of an earlier form of I section having more tapered flanges than the standard universal beams, are also listed. These vary from 203 mm × 102 mm to 76 mm × 76 mm.

Specimen pages from the Handbook are reproduced on pages 49 to 52 by kind permission of The British Constructional Steelwork Association Ltd. and The Constructional Steel Research and Development Organisation. It should be noted that the nominal size is not necessarily the actual size, which is given under the headings D and B. When selecting sections it will be noticed that for a given serial size, there are several actual sizes available. For example if a 356 × 171 UB section was selected, four different actual sizes are available, varying from 364 mm × 173.2 mm to 352 mm × 171 mm. The depth between fillets (d) remains the same, but the thickness of the flange and web varies. With reference to the tables, it can be seen that for a span of 6 m a 457 × 191 × 98 kg Grade 43 UB will support a total uniformily distributed load of 430 kN. A 457 × 191 × 67 kg Grade 43 UB will only support a total load of 285 kN. It is therefore essential when selecting sections to ensure the mass per metre is always specified.

When selecting a universal column from the tables, it will be observed that the B and D dimensions are such, that the cross section of the column is very nearly square. The advantage of this is to restrict the tendency for the column to buckle when axially loaded. For the popular 305 × 305 section there are seven actual sizes available. If the effective length of the column is 3 m the safe concrete load the column can support will vary from a maximum of 5046 kN to a minimum of 1718 kN. Again it is essential that the

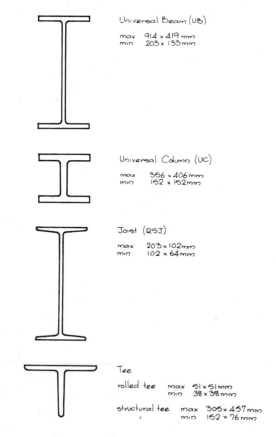

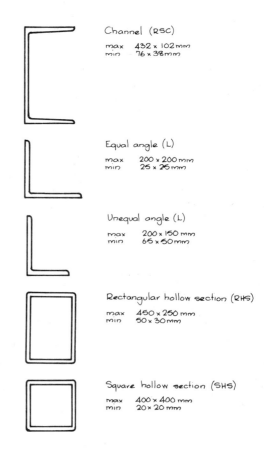

Figure 76 Standard steel sections

mass per metre is stated when selecting column sections.

When specifying angles it is usual to state the leg thickness instead of the mass per metre.

10.2 Methods of connection

Structural steel sections are connected together by the use of bolts or welding. In some cases it is still necessary to use a rivited connection, but only the first two methods are discussed in this book.

Bolts

The most commonly used types are black bolts, and high strength friction grip bolts. Black bolts are classified as grades 4.6 and 8.8 and are specified in BS 4190: 1967 and BS 3692: 1967 respectively. The head and nut must be of hexagonal shape, and tapered washers must be provided for heads and nuts bearing on bevelled surfaces. Holes must be drilled 2 mm or 3 mm larger in diameter than the bolt, depending on its size. The lower strength 4.6 grade is generally used in completing certain types of

connections. Figure 77 gives an example of this type. Grade 8.8 bolts are manufactured more accurately from a higher strength steel. Consequently this type is allowed to resist higher shearing stresses than the grade 4.6 bolt.

High strength friction grip bolts are described in BS 4395 Part 1: 1969. The bolts are manufactured from high tensile steel. Hardened steel washers must

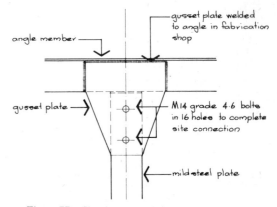

Figure 77 Simple connection

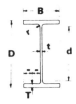

UNIVERSAL BEAMS

DIMENSIONS AND PROPERTIES

Serial size	Mass per metre	Depth of section D	Width of section B	Thickness		Root radius r	Depth between fillets d	Area of section
				Web t	Flange T			
mm	kg	mm	mm	mm	mm	mm	mm	cm²
457 × 191	98	467.4	192.8	11.4	19.6	10.2	407.9	125.3
	89	463.6	192.0	10.6	17.7	10.2	407.9	113.9
	82	460.2	191.3	9.9	16.0	10.2	407.9	104.5
	74	457.2	190.5	9.1	14.5	10.2	407.9	95.0
	67	453.6	189.9	8.5	12.7	10.2	407.9	85.4
457 × 152								
	82	465.1	153.5	10.7	18.9	10.2	406.9	104.5
	74	461.3	152.7	9.9	17.0	10.2	406.9	95.0
	67	457.2	151.9	9.1	15.0	10.2	406.9	85.4
	60	454.7	152.9	8.0	13.3	10.2	407.7	75.9
	52	449.8	152.4	7.6	10.9	10.2	407.7	66.5
406 × 178	74	412.8	179.7	9.7	16.0	10.2	360.5	95.0
	67	409.4	178.8	8.8	14.3	10.2	360.5	85.5
	60	406.4	177.8	7.8	12.8	10.2	360.5	76.0
	54	402.6	177.6	7.6	10.9	10.2	360.5	68.4
406 × 140	46	402.3	142.4	6.9	11.2	10.2	359.6	59.0
	39	397.3	141.8	6.3	8.6	10.2	359.6	49.4
356 × 171	67	364.0	173.2	9.1	15.7	10.2	312.2	85.4
	57	358.6	172.1	8.0	13.0	10.2	312.2	72.2
	51	355.6	171.5	7.3	11.5	10.2	312.2	64.6
	45	352.0	171.0	6.9	9.7	10.2	312.2	57.0
356 × 127	39	352.8	126.0	6.5	10.7	10.2	311.1	49.4
	33	348.5	125.4	5.9	8.5	10.2	311.1	41.8
305 × 165	54	310.9	166.8	7.7	13.7	8.9	265.6	68.4
	46	307.1	165.7	6.7	11.8	8.9	265.6	58.9
	40	303.8	165.1	6.1	10.2	8.9	265.6	51.5
305 × 127	48	310.4	125.2	8.9	14.0	8.9	264.6	60.8
	42	306.6	124.3	8.0	12.1	8.9	264.6	53.2
	37	303.8	123.5	7.2	10.7	8.9	264.6	47.5
305 × 102	33	312.7	102.4	6.6	10.8	7.6	275.8	40.8
	28	308.9	101.9	6.1	8.9	7.6	275.8	36.3
	25	304.8	101.6	5.8	6.8	7.6	275.8	31.4
254 × 146	43	259.6	147.3	7.3	12.7	7.6	218.9	55.1
	37	256.0	146.4	6.4	10.9	7.6	218.9	47.5
	31	251.5	146.1	6.1	8.6	7.6	218.9	40.0
254 × 102	28	260.4	102.1	6.4	10.0	7.6	225.0	36.2
	25	257.0	101.9	6.1	8.4	7.6	225.0	32.2
	22	254.0	101.6	5.8	6.8	7.6	225.0	28.4
203 × 133	30	206.8	133.8	6.3	9.6	7.6	172.3	38.0
	25	203.2	133.4	5.8	7.8	7.6	172.3	32.3

UNIVERSAL BEAMS

DIMENSIONS AND PROPERTIES

Serial size	Moment of inertia			Radius of gyration		Elastic modulus		Ratio $\dfrac{D}{T}$
	Axis x–x		Axis	Axis	Axis	Axis	Axis	
	Gross	Net	y–y	x–x	y–y	x–x	y–y	
mm	cm⁴	cm⁴	cm⁴	cm	cm	cm³	cm³	
457 × 191	45717	40615	2343	19.10	4.33	1956	243.0	23.9
	41021	36456	2086	18.98	4.28	1770	217.4	26.3
	37103	32996	1871	18.84	4.23	1612	195.6	28.8
	33388	29698	1671	18.75	4.19	1461	175.5	31.6
	29401	26190	1452	18.55	4.12	1296	152.9	35.7
457 × 152	36215	32074	1143	18.62	3.31	1557	149.0	24.6
	32435	28744	1012	18.48	3.26	1406	132.5	27.1
	28577	25357	878	18.29	3.21	1250	115.5	30.6
	25464	22611	794	18.31	3.23	1120	103.9	34.2
	21345	19035	645	17.92	3.11	949.0	84.6	41.3
406 × 178	27329	24062	1545	16.96	4.03	1324	172.0	25.9
	24329	21425	1365	16.87	4.00	1188	152.7	28.6
	21508	18934	1199	16.82	3.97	1058	134.8	31.8
	18626	16457	1017	16.50	3.85	925.3	114.5	37.0
406 × 140	15647	13765	539	16.29	3.02	777.8	75.7	36.0
	12452	11017	411	15.88	2.89	626.9	58.0	46.0
356 × 171	19522	17045	1362	15.12	3.99	1073	157.3	23.2
	16077	14053	1109	14.92	3.92	896.5	128.9	27.5
	14156	12384	968	14.80	3.87	796.2	112.9	30.9
	12091	10609	812	14.57	3.78	686.9	95.0	36.2
356 × 127	10087	9213	357	14.29	2.69	571.8	56.6	33.1
	8200	7511	280	14.00	2.59	470.6	44.7	41.0
305 × 165	11710	10134	1061	13.09	3.94	753.3	127.3	22.7
	9948	8609	897	13.00	3.90	647.9	108.3	26.0
	8523	7384	763	12.86	3.85	561.2	92.4	29.9
305 × 127	9504	8643	460	12.50	2.75	612.4	73.5	22.2
	8143	7409	388	12.37	2.70	531.2	62.5	25.4
	7162	6519	337	12.28	2.67	471.5	54.6	28.4
305 × 102	6487	5800	193	12.46	2.15	415.0	37.8	29.0
	5421	4862	157	12.22	2.08	351.0	30.8	34.8
	4387	3962	120	11.82	1.96	287.9	23.6	44.6
254 × 146	6558	5706	677	10.91	3.51	505.3	92.0	20.4
	5556	4834	571	10.82	3.47	434.0	78.1	23.4
	4439	3879	449	10.53	3.35	353.1	61.5	29.1
254 × 102	4008	3569	178	10.52	2.22	307.9	34.9	26.0
	3408	3046	148	10.29	2.14	265.2	29.0	30.8
	2867	2575	120	10.04	2.05	225.7	23.6	37.2
203 × 133	2887	2476	384	8.72	3.18	279.3	57.4	21.5
	2356	2027	310	8.54	3.10	231.9	46.4	26.0

In calculating the net moment of inertia, each flange
of 300 mm or greater width is reduced by two holes,
and each flange less than 300 mm wide by one hole.

UNIVERSAL BEAMS

SAFE LOADS FOR GRADE 43 STEEL

Serial size	Mass per metre	Safe distributed loads in kilonewtons for spans in metres and deflection coefficients													Critical span Lc
		2.00	2.50	3.00	3.50	4.00	4.50	5.00	5.50	6.00	7.00	8.00	9.00	10.00	
mm	kg	112.0	71.68	49.78	36.57	28.00	22.12	17.92	14.81	12.44	9.143	7.000	5.531	4.480	m
457 × 191	98	†1066	1033	861	738	645	574	516	469	430	369	323	287	258	4.00
	89	†983	935	779	668	584	519	467	425	389	334	292	260	234	3.86
	82	†911	851	709	608	532	473	426	387	355	304	266	236	213	3.74
	74	†832	771	643	551	482	429	386	351	321	276	241	214	193	3.65
	67	†771	684	570	489	428	380	342	311	285	244	214	190	171	3.53
457 × 152	82	†995	822	685	587	514	457	411	374	343	294	257	228	206	3.03
	74	†913	742	619	530	464	412	371	337	309	265	282	206	186	2.92
	67	825	660	550	471	413	367	330	300	275	236	206	183	165	2.81
	60	†728	591	493	422	370	329	296	269	246	211	185	164	148	2.78
	52	626	501	418	358	313	278	251	228	209	179	157	139	125	2.62
406 × 178	74	†801	699	583	499	437	388	350	318	291	250	218	194	175	3.65
	67	†721	627	523	448	392	348	314	285	261	224	196	174	157	3.54
	60	†634	559	466	399	349	310	279	254	233	200	175	155	140	3.45
	54	611	489	407	349	305	271	244	222	204	174	153	136	122	3.28
406 × 140	46	513	411	342	293	257	228	205	187	171	147	128	114	103	2.58
	39	414	331	276	236	207	184	165	150	138	118	103	92	83	2.41
356 × 171	67	†662	567	472	405	354	315	283	258	236	202	177	157	142	3.72
	57	†574	473	394	338	296	263	237	215	197	169	148	131	118	3.50
	51	†519	420	350	300	263	234	210	191	175	150	131	117	105	3.38
	45	453	363	302	259	227	201	181	165	151	130	113	101	91	3.23
356 × 127	39	377	302	252	216	189	168	151	137	126	108	94	84	75	2.33
	33	311	248	207	177	155	138	124	113	104	89	78	69	62	2.18
305 × 165	54	†479	398	331	284	249	221	199	181	166	142	124	110	99	3.69
	46	†412	342	285	244	214	190	171	155	143	122	107	95	86	3.53
	40	370	296	247	212	185	165	148	135	123	106	93	82	74	3.38
305 × 127	48	404	323	269	231	202	180	162	147	135	115	101	90	81	2.59
	42	351	280	234	200	175	156	140	127	117	100	88	78	70	2.45
	37	311	249	207	178	156	138	124	113	104	89	78	69	62	2.37
305 × 102	33	274	219	183	156	137	122	110	100	91	78	68	61	55	1.90
	28	232	185	154	132	116	103	93	84	77	66	58	51	46	1.79
	25	190	152	127	109	95	84	76	69	63	54	47	42	38	1.64
254 × 146	43	333	267	222	191	167	148	133	121	111	95	83			3.41
	37	286	229	191	164	143	127	115	104	95	82	72			3.22
	31	233	186	155	133	117	104	93	85	78	67	58			2.96
254 × 102	28	203	163	135	116	102	90	81	74	68	58	51			2.01
	25	175	140	117	100	88	78	70	64	58	50	37			1.87
	22	149	119	99	85	74	66	60	54	50	43	37			1.75
203 × 133	30	184	147	123	105	92	82	74	67	61	53				3.03
	25	153	122	102	87	77	68	61	56	51					2.80

Loads printed in italic type do not cause overloading of the unstiffened web, and do not cause deflection exceeding span/360.

Loads printed in ordinary type should be checked for deflection

† Load is based on allowable shear of web and is less than allowable load in bending.

UNIVERSAL COLUMNS
AS STANCHIONS

SAFE LOADS FOR GRADE 43 STEEL

Serial size	Mass per metre	Safe concentric loads in kilonewtons for effective lengths in metres												
mm	kg	1.5	2.0	2.5	3.0	3.5	4.0	5.0	6.0	8.0	10.0	12.0	14.0	16.0
356 × 406	634	**11313**	**11313**	**11313**	**11313**	**11313**	**11313**	11002	10517	9090	7287	5643	4382	3462
	551	**9826**	**9826**	**9826**	**9826**	**9826**	**9826**	9527	9087	7801	6206	4782	3705	2923
	467	**8337**	**8337**	**8337**	**8337**	**8337**	**8337**	8057	7668	6534	5158	3955	3056	2408
	393	**7012**	**7012**	**7012**	**7012**	**7012**	7001	6755	6414	5427	4253	3247	2503	1970
	340	**6057**	**6057**	**6057**	**6057**	**6057**	6040	5821	5518	4644	3620	2755	2121	1668
	287	5455	5382	5309	5237	5174	5102	4911	4647	3889	3015	2288	1759	1381
	235	4467	4406	4346	4286	4235	4174	4013	3790	3153	2432	1840	1412	1108
Column Core	477	8501	8501	8501	8501	8501	8494	8201	7795	6615	5200	3978	3070	2417
356 × 368	202	3832	3778	3723	3671	3621	3560	3398	3173	2555	1917	1430	1090	852
	177	3353	3305	3257	3211	3168	3114	2970	2770	2224	1665	1241	945	739
	153	2899	2857	2815	2776	2737	2690	2563	2388	1909	1426	1061	807	631
	129	2448	2413	2377	2344	2311	2270	2161	2010	1601	1192	885	673	526
305 × 305	283	**5046**	**5046**	**5046**	**5046**	4964	4840	4505	4048	2981	2116	1539	1158	
	240	4507	4431	4356	4288	4200	4091	3796	3397	2480	1753	1272	957	
	198	3718	3654	3593	3534	3459	3366	3113	2772	2006	1412	1023	768	
	158	2962	2911	2862	2813	2751	2673	2462	2180	1562	1095	792	594	
	137	2570	2524	2481	2438	2383	2314	2126	1877	1338	935	676	507	
	118	2204	2164	2127	2090	2041	1981	1816	1599	1134	791	571		
	97	1813	1780	1749	1718	1677	1626	1487	1304	920	640	461		
254 × 254	167	3100	3036	2980	2905	2809	2686	2363	1977	1307	886	632		
	132	2445	2394	2348	2286	2206	2104	1838	1526	1000	676			
	107	1991	1949	1910	1858	1790	1704	1480	1222	796	537			
	89	1659	1624	1592	1547	1490	1417	1227	1010	656	442			
	73	1351	1323	1295	1259	1210	1149	991	813	525	354			
203 × 203	86	1580	1541	1488	1416	1321	1205	950	728	443				
	71	1306	1273	1229	1168	1087	990	777	594	360				
	60	1086	1058	1020	966	896	812	632	481	291				
	52	951	927	892	845	783	708	550	418	252				
	46	842	820	789	745	689	622	481	365	220				
152 × 152	37	661	626	574	505	429	357	249	180					
	30	533	504	460	403	341	283	197	142					
	23	413	388	351	303	253	208	144	103					

The above safe loads are tabulated for ratios of slenderness not exceeding 180. Values in ordinary type are calculated in accordance with Table 17a of BS 449. Values in bold type are based on the limiting stress of 140 N/mm² for material over 40 mm thick.

be used under the head and nut. The bolts are tightened up to a predetermined extent and the loads are then transferred from one member to another by friction instead of the shear and bearing strengths of the bolt. Rules for the use of this type of bolt are given in BS 4604 Part 1: 1970.

Welding

The advantage of this type of connection is that the joints between members are secured with greater rigidity, and a neater connection is possible. On the debit side, it is necessary to adopt expensive procedures, such as an X-ray to ensure that the weld is of an acceptable standard and strength.

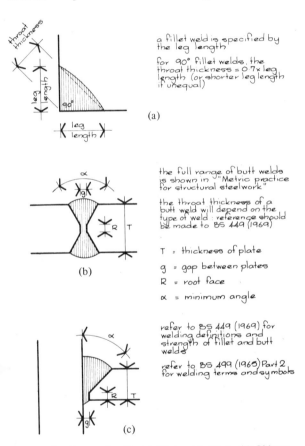

Figure 78   Types of weld: (a) fillet weld; (b) double V butt weld; (c) single bevel butt weld

Butt and fillet welds are the two most common types used in structural steelwork. The connection between members is made using the electric-arc welding process. All welding should be carried out in accordance with the requirements of BS 5135: 1974. Details of the two types of welds are shown in Figure 78.

10.3 Site connections

It is preferable for welded connections to be made in the fabrication shop and the connection completed on site by the use of bolts. Site welding should be avoided if possible, as this could prove to be an expensive operation. Welding equipment will have to be brought to the site and possibly lifted to the welding location. It will also be difficult for the site weld to be tested for strength.

Beam to column connection

Figure 79 illustrates a simple connection where the beam loading is transmitted to the column via a seating angle cleat, which has been fillet welded to the column in the fabrication shop. A top angle cleat is fillet welded to the beam during fabrication, and the connection is completed by site bolting.

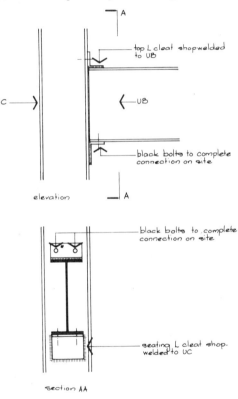

Figure 79   Beam to column connection (welding and bolting)

Column to baseplate connection

There are two methods of connecting a column to a baseplate. These are known as:

(i) Slab or Bloom;
(ii) Gusset.

53

The object in both cases is to spread the column load on to the concrete foundation. It is essential in case (i) for the end of the column to be machined dead square, to ensure a good interface contact with the baseeplate. It is also essential that the plate is made thick enough to distribute the column load as necessary.

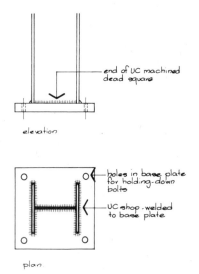

elevation

plan

Figure 80 Slab base

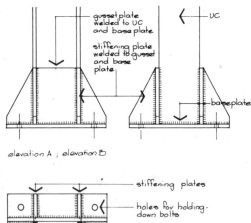

elevation A ; elevation B

plan

A

Figure 81 Gusseted base

The column to baseplate connection can be made by fillet welding, see Figure 80. The gusset type is useful when it is necessary to transmit a high bending moment to the foundation. As a gusset or stiffening plate is used, the thickness of baseplate can be reduced. Figure 81 shows a typical detail. Holding down bolts must be provided in both cases. The bolts must be designed to resist the tendency for the column to overturn.

Column splices

When it is necessary to join column lengths together, the join or splice is made between 450 mm and 600 mm above floor level so that the beam connections

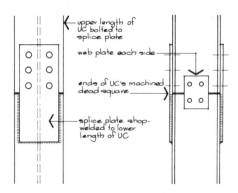

Figure 82 Column splices with web plate

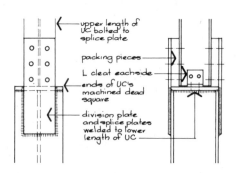

Figure 83 Column splice with packing pieces

are not obstructed. Any change in the column section is made at this point. It is essential that the end of the columns are machined dead square to ensure that the load is transmitted uniformily to the lower column. Figures 82 and 83 illustrate different types of column splices.

Beam to beam connection

It is sometimes necessary for a secondary beam to be connected to a main beam. This can be achieved by fillet welding an end plate to the secondary beam

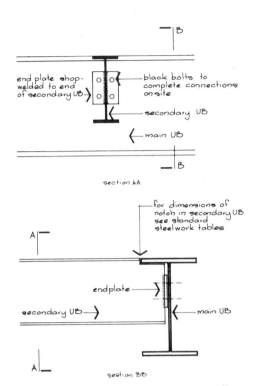

section AA

for dimensions of notch in secondary UB see standard steelwork tables

section BB

Figure 84 Beam to beam connection (welding and bolting)

during fabrication and completing the connection by site bolting. If the top flange of the beams are level, it will be usually necessary to notch the flange of the secondary beam, see Figure 84.

10.4 Erection procedure

The space available to the steelwork erector will largely govern the selection of the erection plant. Some of the points to consider when selecting plant for any particular construction work will be:

(i) Proposed method of erection;
(ii) Speed of erection;
(iii) Height of the structure;
(iv) Reach required of the plant.

If a two-storey high structure is considered, and providing conditions permit, then a mobile crane could be ideal. It can commence work immediately upon its delivery to the site, and when erection is completed it can be dispatched to another site within a very short time. When selecting the crane, attention should be paid to the weight and length of the structural members, so that economic use can be made of the crane. The steelwork should be

erected in bays from ground level to roof, before the next bay is commenced.

If the site is congested and the structure is several storeys high, then a tower crane may well prove to be economic. The crane will probably be used for other purposes, such as the placing of the floor slabs and erecting the cladding, etc. When storage space is limited it is essential that only the steel for each section is delivered to the site. The steelwork fabricator must carefully schedule the steel so that it arrives at the correct rate for erection purposes.

10.5 Fire protection

In the event of a fire, the time for a steel frame to attain a condition where it is no longer capable of sustaining its design load depends not only on the physical size of the steel sections, but also on the extent to which the steelwork is protected by insulating material.

The fire protection of structural steelwork is essential, although there are certain exemptions. Reference should be made to the Building Regulations for more information on these exemptions. Structural steelwork can be protected by one of several methods. These are described in the following paragraphs.

(i) Solid protection

With this method it is necessary to encase the steel members with *in situ* concrete suitably reinforced, see Figure 85. The operation can be expensive, due to the cost of additional formwork. Time consuming, as the concrete protection must be allowed to harden, and add unnecessary dead-weight to the structure.

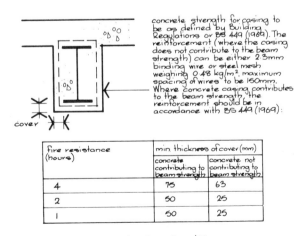

cover

concrete strength for casing to be as defined by Building Regulations or BS 449 (1969). The reinforcement (where the casing does not contribute to the beam strength) can be either 2.3mm binding wire or steel mesh weighing 0.48 kg/m², maximum spacing of wires to be 150mm. Where concrete casing contributes to the beam strength, the reinforcement should be in accordance with BS 449 (1969):

fire resistance (hours)	min. thickness of cover (mm)	
	concrete contributing to beam strength	concrete not contributing to beam strength
4	75	63
2	50	25
1	50	25

Figure 85 Solid protection for universal beam

(ii) Hollow protection

The advantage of this method is that the weight of the cladding can be reduced, and the progress of the work can be speeded up, by the use of vermiculite insulating board. These boards are manufactured in various thicknesses to suit different fire periods, see Figure 86.

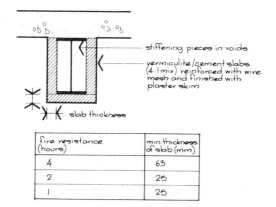

fire resistance (hours)	min thickness of slab (mm)
4	63
2	25
1	25

Figure 86 Hollow protection for universal beam

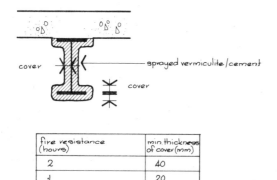

fire resistance (hours)	min thickness of cover (mm)
2	40
1	20

Figure 87 Profile protection for universal beam

Alternatively expanded metal lathing coated with gypsum plaster may be used. It should be noted that this method is not necessarily cheaper than the solid protection.

(iii) Profile protection

In this case the protecting material, which can be vermiculite, is sprayed direct to the steel section. A thickness of 20 mm will give a fire protection of 1 hour. The disadvantage of this method is its untidy appearance. Figure 87 shows a typical section through a steel member coated with vermiculite.

10.6 Reinforced concrete

Plain concrete has considerable strength in compression, but very little strength in tension. The ratios vary between 10 : 1 and 15 : 1. As the tensile strength is low, it follows that concrete is also weak in bending and shear. The tensile strength is also unreliable, it may be entirely destroyed by shock or sudden jar, or as the result of shrinkage. The application of plain concrete is normally limited to uses when compressive strength and mass are the principal requirements such as gravity retaining walls.

In most structural cases, high tensile stresses have to be accommodated. For this purpose steel reinforcement is embedded in the concrete at the time of casting, so forming the composite material known as reinforced concrete. The reinforcement, which may be either mild, or high tensile, has considerable strength in both tension and compression. By careful design the reinforcement can be so disposed in the concrete as to be available to take all the tensile stresses wherever they occur.

The design of reinforced concrete is based on the limit state concept as related to CP 110: 1972. It is impossible to arrive by calculation, at every precise margin there may be between the failure strength of a reinforced concrete member, and the load to which that member is likely to be subjected at any time in its life. This is partly due to the uncertainty of the

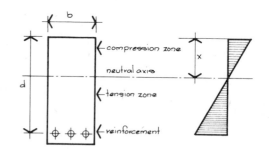

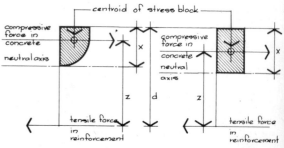

note: when the simplified method is used, the maximum value of x = ½d, and the minimum value of z = ¾d

Figure 88 (a) cross section through beam; (b) strain diagram; (c) true parabolic stress block; (d) simplified rectangular stress block

maximum loads that will be actually applied to the member, and partly due to the variations in the strength properties of the concrete and steel in the member.

To allow for possible variations in workmanship of quality control a partial safety factor is employed for both steel and concrete. These values are generally 1.15 and 1.5 respectively. The partial safety factors for loading varies and CP 110: 1972 should be consulted for the required values.

As mentioned previously concrete is strong in compression and weak in tension. For calculation purposes the concrete in the tension zone is ignored. Figure 88 (c) shows the stress block from which the design tables in CP 110: 1972, Part 2 have been based. For a truly balanced section the force in compression must equal the force in tension. In practice, it is rarely possible for the two values to be balanced. It is normal for the concrete or steel to be slightly under stressed.

In symbols the ultimate resistance moments for the beam are:

$$Mu \text{ (concrete)} = 0.15 \, fcu \, bd^2$$
$$Mu \text{ (steel)} = 0.65 \, fy \, As \, d$$

These values, which are listed in CP 110 are based on the simplified stress block which is shown in Figure 88 (d). Where:

fcu = Characteristic concrete cube strength.
fy = Characteristic strength of reinforcement.
b = Width of section.
d = Effective depth of tension reinforcement.
As = Area of tension reinforcement.

It can be seen that the factors which affect the design are:

Neutral axis depth (x)
Effective depth of tension reinforcement (d)
Lever arm (z)
Width of section (b)

The neutral axis and the lever arm values will vary with differing widths and dpeths of the beam section. It therefore follows that if the reinforcement is not correctly positioned on site, the basis of the design is upset. This can result in overstressing of the steel or concrete.

An inspection of works should ensure that the beam, slab, etc. is of correct size and shape, and that the reinforcement is fixed in accordance with the engineer's drawings. Particular attention should be given to the positioning of the tension reinforcement.

(Several instances have been reported of complete failures of concrete balconies due to the fact that the cantilever reinforcement has been positioned incorrectly).

All reinforcement must be surrounded by concrete to provide protection against corrosion, fire etc and to allow the bar to develop its correct force. CP 110: 1972 relates the cover of the reinforcement to the grade of concrete and condition of exposure. The nominal cover should never be less than the diameter of the bar. An extract from CP 110: 1972 for the normal range of conditions is given in Table 10.

Table 10 Nominal concrete cover to reinforcement for concrete grades (*From CP 110: 1972*)

Condition of exposure	*Nominal cover (mm) for concrete grade of*			
	25	*30*	*40*	*50*
Mild	20	15	15	15
Moderate	40	30	25	20
Severe	50	40	30	25

In the heavier loaded beams, links are provided to tie the reinforcement together, and to also minimise the shrinkage cracks. These links should be positioned in accordance with the engineer's drawings. It will be noticed that in some cases the centres of the links are closed up towards the ends of the beam. This is to resist the shear force in the beam, which may be critical at that point; see Figure 89.

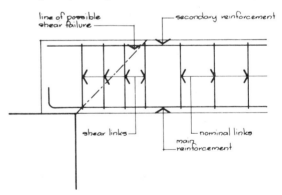

Figure 89 Position of shear links in simple beam

10.7 Reinforced concrete details

Column to base connection

In this case reinforcement is placed in both directions in the bottom of the base slab. This is to resist the

Standard detailing notation: 11-Y20-1-200 (B) denotes
11 no. high yield 20mm diameter
bars bar mark 1 at 200mm
centres in bottom of base

note: where 'R' is indicated this denotes mild steel reinforcement

the cover to reinforcement
will depend on the grade
of concrete used.
reference should be
made to the appropriate
table in CP 110 (1972)

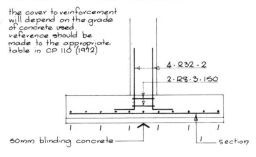

4-R32-2
2-R8-3-150

50mm blinding concrete —————— Section

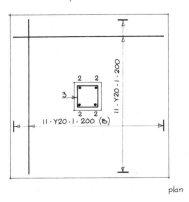

11-Y20-1-200
2   2
3
2   2
11-Y20-1-200 (B)

plan

Figure 90   Column to square base connection

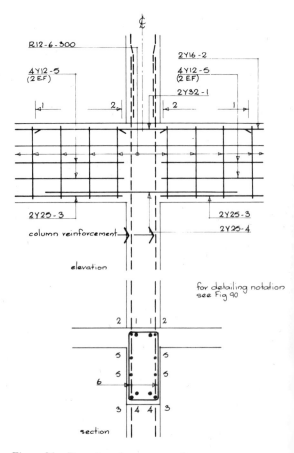

R12-6-300

2Y16-2

4Y12-5
(2 EF)

4Y12-5
(2 EF)

2Y32-1

2Y25-3

2Y25-3
2Y25-4

column reinforcement

elevation

for detailing notation
see Fig 90

section

Figure 91   Beam to column connection

tensile bending stresses which have been set up in the base. The starter bars for the column are supported on the base reinforcement, and links are provided to stabilise and locate the starter bars during the construction process. Figure 90 shows a typical detail.

Beam to column

An illustration for a beam to column connection is shown in Figure 91. The reinforcement in the beam should be carefully arranged so that it does not coincide with the column reinforcement.

Precast concrete beam to in-situ reinforced concrete floor

It is usual for this type of beam to be designed so that it acts intergrally with the *in-situ* concrete floor. It is therefore necessary for some form of connector to be provided to enable the beam and slab to act together.

The usual method is to provide mild steel links which have been left projecting from the top of the beam. The top steel reinforcement in the slab is wired to these links. When the slab is cast, it will enable the beam and slab to act as an integral unit, see Figure 92.

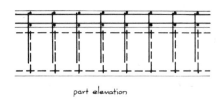

part elevation

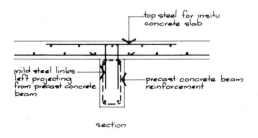

top steel for insitu
concrete slab

mild steel links
left projecting
from precast concrete
beam

precast concrete beam
reinforcement

section

Figure 92   Precast concrete beam *in situ* R.C. slab connection

58

11. PRINCIPLES AND CONSTRUCTION OF FORMWORK

11.1 Requirements

To be successful in its function, formwork must fulfil the following requirements.

(i) It should be built accurately, so that the desired size, shape, position and finish is attained.
(ii) The formwork should be strong enough to support all dead and live loads without collapse, or danger to site personnel, and to the structure.
(iii) The choice of the type of formwork material will depend on the amount of repetition anticipated. Standardisation of forms should be aimed at as far as possible.
(iv) The joints in the formwork should be grout tight as far as possible.
(v) The formwork should be so designed that it can be erected and dismantled easily.

11.2 Materials used for formwork

The material used for formwork depends on the type of structure to be erected. It can be either timber, steel or plastic.

Frequently the choice of timber is a question of local availability and cost. Any timber that is straight and structurally strong and sound, may be used for formwork. Partially seasoned stock is generally used, since fully dried timber swells excessively when it becomes wet. Unseasoned timber will dry out and warp during hot weather. It should be noted that old and new boards should not be used together in the same panel.

Plywood has been increasingly used for formwork panels in recent years, as it is light, strong, and can be supplied in sheets up to 3000 mm in length by 1200 mm in width. The plywood should be exterior grade and the thickness of the sheets should be related to the anticipated pressures that the panels will be subjected.

With reasonable care, steel forms will last a considerable time, and hence will be suitable for many reuses. The selection may depend on cost comparison with other materials, but more likely, it will be used where a large amount of repetitive work is possible. This material is ideal when a modular prefabricated panel system is used. These panels are manufactured in small easily handled units, and can be either purchased or rented from a manufacture. Further information on system formwork can be obtained from manufacturers such as Acrow (Engineers) Ltd. or Kwikform Ltd.

Glass fibre or plastic forms are finding favour in concrete construction, because it produces an excellent concrete finish. As the materials used require the temperature and humidity to be carefully controlled, the formwork must be manufactured under factory conditions. This may limit its use to complex shaped forms which can be manufactured off site, although some standard arrangements are now available.

Glass fibre or plastic moulds are now used in reinforced concrete floor construction. When the moulds are removed they leave a honeycomb or waffle effect on the underside of the floor.

11.3 Supports for formwork

Beam and slab forms are usually supported by timber or adjustable steel props, although in some cases tubular steel scaffolding is used. It is essential that the props are adequately braced, and can support all dead and live loads safely.

11.4 Treatment and finishes to formwork

Quite often a concrete finish is required that is superior to that obtained by the use of timber boarding. The latter frequently leaves a ridge or 'fin' where joints occur. To achieve this superior finish it is common to use plywood, steel or an oil impregnated hardboard as a lining to the framework, although plastic sheeting has been successfully used. Factory produced precast concrete units are also used as a facing for reinforced concrete sections.

To prevent concrete from sticking to the surface when the formwork is removed, and thus spoiling the concrete finish, it is necessary for the forms to be treated with a release agent. Several types are available, but neat oil with surfactant, or mould cream emulsions appear to be most popular. It is important that the release agent used should not stain the concrete surface, particularly if a fair faced finish is required. Oil should be used sparingly and should not be allowed to come into contact with the reinforcement.

Formwork must be watertight to prevent the loss of grout, which can cause staining of the concrete, and in some cases honeycombing. The joints in the forms should be sealed either by the use of sealing strips or waterproof adhesive tape.

11.5 Striking of formwork

The contractor has general responsibility for the design, construction and safety of the formwork. The time for the removal or striking of the forms should be specified by the engineer. It is clearly economic to remove the formwork as early as possible, although the time-scale should relate to the concrete strength and the time of year that the concrete has been placed.

Table 11 Minimum periods before striking formwork

Formwork	Surface temperature of concrete	
	16°C	7°C
Vertical forms to columns, walls and large beams	9 hours	12 hours
Slab soffits (props left under)	4 days	7 days
Beam soffits	8 days	14 days
Props to slabs	11 days	14 days
Props to beams	15 days	21 days

A general guide for the removal of forms is given in CP 110: 1972 and is shown in Table 11.

In very cold weather, it is recommended that the periods given above should be increased as follows:

(i) For each day that the temperature is below 7°C, add half a day to the period given in the right-hand column.

(ii) For each day that the temperature is below 2°C add one day to the period given in the right hand column.

Formwork should be removed carefully and slowly to avoid damage to the partially hardened concrete. The careful removal will also prolong the life of the forms.

12. PRINCIPLES AND CONSTRUCTION OF TRUSSES IN TIMBER AND STEEL

12.1 Necessity for roof trusses

As described in the TEC Unit Construction Technology I (U.75/073), the important function of a roof is to provide protection to the structure from the weather. BS 3589: 1963 defines a 'pitched roof' as any roof whose angle of slope to the horizontal lies between 10° and 70°. If the angle of slope is below 10° then the roof is defined as 'flat'.

Roof structures are classified in terms of span as:

Short span up to 7.50 m.
Medium span 7.50 m to 24.50 m.
Long span over 24.50 m.

Roofs in the short span range can be either 'flat' or 'pitched' and are usually constructed in timber. Medium span roofs are usually fabricated from standard steel sections and are classified either as a truss or a lattice girder. Figures 93 and 94 show an arrangement of both types. Long span roof

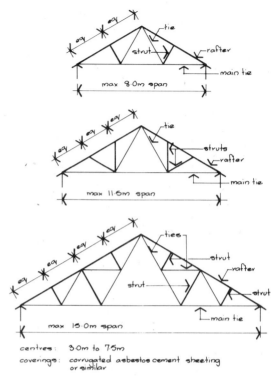

centres: 3.0m to 7.5m
coverings: corrugated asbestos cement sheeting or similar

Figure 93 Standard roof trusses

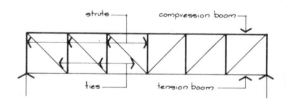

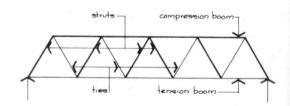

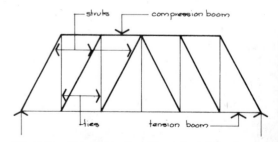

Figure 94 Standard lattice girders: (top) 'N' or Pratt girder; (centre) Warren girder; (bottom) Howe girder

construction is highly specialised, and can involve space decks and girder construction. These types are outside the scope of this book.

The decision as to the type of roof to be used will largely be influenced by the span of the structure to be covered, by the roof coverings and loading, or by architecture considerations.

The principle of roof truss or lattice girder design and construction, is that the loading from the roof coverings is transferred to the purlins, which in turn transmit the load to the roof structure. It will be observed from Figures 93 and 94 that the members are framed up in such a way so as to produce a triangulated structure. The intersections of these members are known as the 'node' points. It will be noted that the loading from the purlins are applied at these points. If the principle of triangulation is observed, then these loads will induce direct force only in the members, see Figure 95.

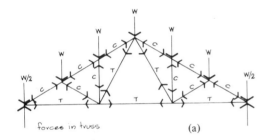

forces in truss   (a)

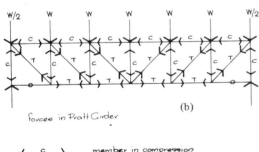

forces in Pratt Girder   (b)

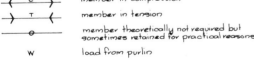

Figure 95   Forces in roof members: (a) forces in truss; (b) forces in Pratt girder

This arrangement is not always practicable, as the purlin spacings vary with the type of roof coverings being used. It is not unusual for purlins to be supported between 'node' points. In this case the rafter members have to be designed for local bending, as well as direct force.

12.2 Steel truss

Figures 97 and 98 show details of a sheet angle truss, with bolted or alternatively welded connections. As discussed in 12.1 the members are set out so that their centres of gravity meet at the 'node' points. This has been followed as far as the welded truss is concerned.

The bolted truss has been set out so that the bolt lines intersect at the 'node' points. The gusset plate must be thick enough and shaped correctly, to resist the forces in the members at the joints. Figure 99 shows a typical form of covering for the above trusses.

12.3 Timber truss

For many years the Timber Research and Development Association (TRADA) have prepared standard designs for timber roof trusses suitable for domestic and industrial use. The trusses span from a minimum of 5.10 m to a maximum of 20.10 m.

The roof coverings vary from tiles to asbestos cement sheeting depending on the type of truss. Detail of a typical industrial truss in shown in Figure 100.

12.4 Timber connections

Timber trusses are fabricated in several ways, they include bolted and connector, nailed and glued joints or a combination of two types. The two most commonly used connectors are the toothed plate and the split ring, which is illustrated in Figure 96. The advantage of the split-ring type is it can carry a greater load, but on the debit side, it requires that the timbers to be jointed are accurately machine grooved. It is usual for trusses with this type of connection to be fabricated under factory conditions.

The toothed plate connector is useful when it is necessary to join sawn timbers, as the joint does not require the same degree of accuracy. In both cases the joint is completed by the addition of a bolt and washers. The size of the bolt will depend on the load to be resisted.

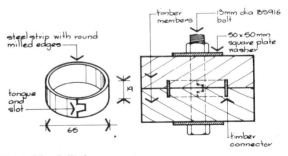

Figure 96   Split ring connector

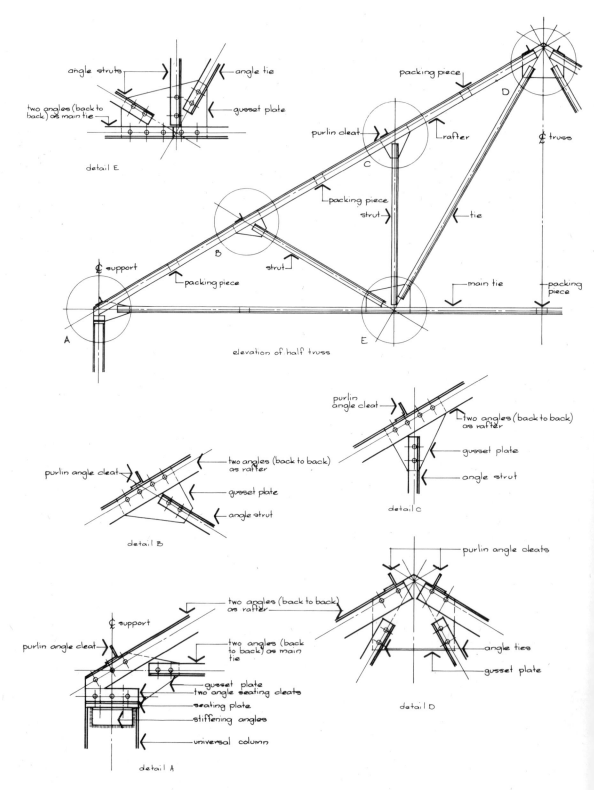

Figure 97 Details of bolted roof truss

62

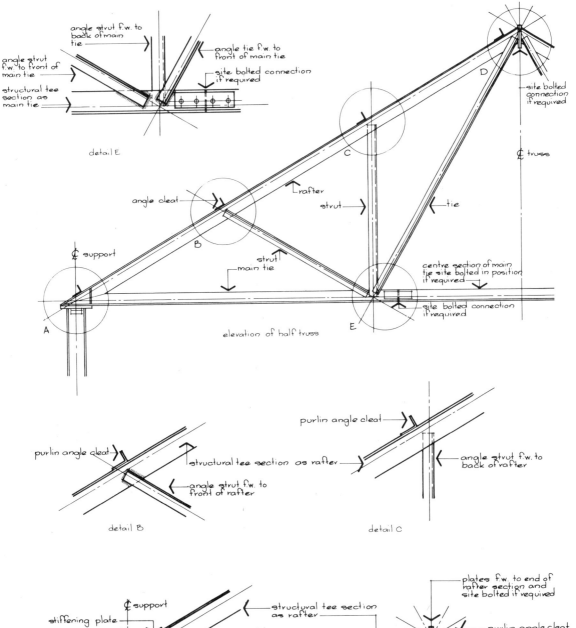

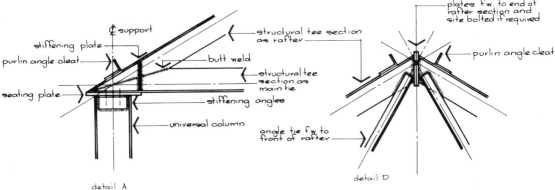

Figure 98 Details of welded roof truss

63

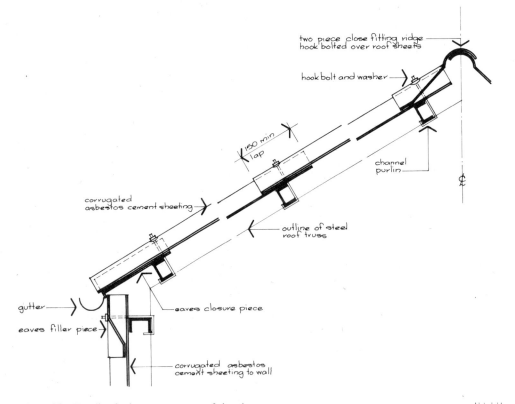

Figure 99 Details of asbestos cement roof sheeting

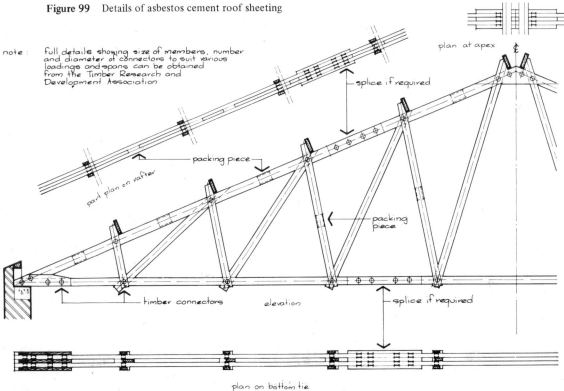

note : full details showing size of members, number
and diameter of connectors to suit various
loadings and spans can be obtained
from the Timber Research and
Development Association

Figure 100 Typical TRADA industrial truss

13. METHODS OF INFILLING FRAMES

The methods of infilling frames are listed below. It is considered that these specific objectives are well covered by existing textbooks. They should be classified within the scope of building technology rather than civil engineering technology.

1. Construction of non-load bearing brick panels including methods of attachment to concrete

2. Construction of timber framed infill panels with plywood and asbestolux external cladding including insulation

3. Steel and aluminium windows including fixing to concrete

4. Principle of patent glazing

5. Details of aluminium patent glazing

Further reading

Readers who require a more in depth knowledge of some of the objectives in Topic Area C may find the following reading list useful.

C.8   Foster, Jack Stroud, and Harrington, Raymond, *Mitchell's building construction. Structure and fabric: Part 2*, B. T. Batsford Ltd (1976).

C.9. *Concrete practice*, Cement and Concrete Association (1975).

C.10. *Structural steelwork handbook*, The British Constructional Steelwork Association (1978).

Metric practice for structural steelwork, The British Constructional Steelwork Association.

Leech, L. V. *Structural steelwork for students*, Newnes-Butterworths (1972).

Boughton, Brian, *Reinforced concrete detailer's Manual*, Crosby Lockwood Staples (1975).

Standard method of detailing reinforced concrete, The Concrete Society.

Astill, A. W. and Martin, H. *Elementary structural design in concrete to CP 110*, Edward Arnold (1975).

Clements, R. W., *Design of reinforced concrete elements*, Macdonald and Evans Ltd (1977).

Chudley, R., *Construction technology*, Vol 3, Longman Group Ltd (1976).

C.11. McKay, J. K., *Building construction Vol 4*, Longman Group Ltd (1975)

C.12. *Design in timber*, Timber Research and Development Association.

McKay, J. K., *Building Construction, Vol 4*, Longman Group Ltd (1975).

C.13. Barry, R., *The construction of building 4*, Crosby Lockwood Staples (1974).

Topic Area D – Internal Construction

14. CONSTRUCTION OF TIMBER, R.C. AND PRECAST CONCRETE SUSPENDED FLOORS

14.1 Suspended floors

Suspended floors may be constructed in timber, reinforced concrete or precast concrete. The choice of floor type will depend on a number of factors. These include: load carrying capacity, fire resistance, sound insulation and economic span.

For the small structure the choice of floor is usually decided by the requirements of loading, span, cost, sound insulation, and speed of erection. In this case a timber floor may well prove to be suitable especially if the superimposed load is small. It should be noted that as the fire resistance is low and the sound insulation is less than a concrete floor the use of timber is usually limited to low rise domestic structures.

With a multi-storey structure other factors need to be considered when deciding on the type of floor to be used. The choice may depend on the type of structural frame, the degree of fire resistance to be provided and the sound insulation required. In this case an *in situ* reinforced concrete floor or precast concrete units may well prove to be the most suitable.

14.2 Timber floors

When discussing timber floors there are two types to be considered.

(i) Single floor

With this type of floor construction, common joists span the whole distance from support to support. The safe span limit for single floors is usually 4.5 m to 5.0 m. The size and spacing of joists depend on the span and loading to be supported and are defined in the Building Regulations. However the size and spacing of joists may be obtained by calculation.

When the span of the joists exceeds about 2.5 m it is necessary to provide some form of lateral restraint; this can be provided at mid span by herringbone strutting which greatly increases the rigidity of the floor.

The ends of the joists can be supported either, by building into the wall (providing the joists are treated

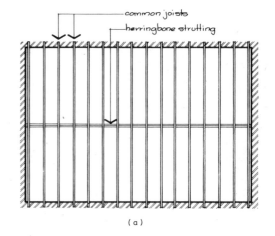
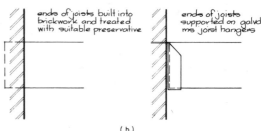
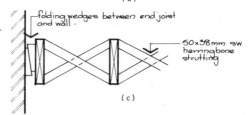

Figure 101 Single timber floor: (a) plan; (b) alternative methods of supporting joists; (c) detail of lateral restraint

with a suitable preservation) or by supporting them on galvanised mild steel joist hangers.

Details of a single timber floor are shown in Figure 101.

(ii) Double floor

When the span of a single timber floor exceeds about 4.5 m it may well prove to be economic to provide intermediate beams, which are usually steel universal sections or precast concrete beams spanning between supports.

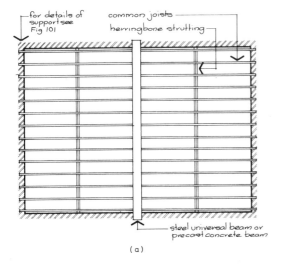

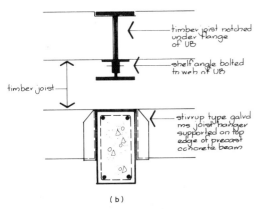

Figure 102 Double timber floor: (a) plan; (b) alternative methods of supporting timber joists

These intermediate beams are so positioned that the span of the common joists are limited from 4.5 to 5.0 m.

Details of this type of floor are shown in Figure 102.

The finish for timber floors usually consists of 1200 × 600 × 19 mm thick flooring grade chipboard nailed all around. Tongued-and-grooved softwood boarding of various thicknesses and widths may also be used.

14.3 Reinforced concrete floors

A simple *in situ* reinforced concrete floor slab consists of main reinforcement spanning one way between supports. Distribution bars laid at right angles and wired to the main bars are provided. The reinforcement used is generally of the high yield deformed type although a mesh reinforcement is sometimes found to be more economic.

Concrete has a greater fire resistance and better sound insulating properties than timber, although the deadweight of the concrete floor is much greater. Details of a reinforced concrete slab is shown in Figure 103.

The desirability of reducing the deadweight of the concrete floor to a minimum for reasons of economy has resulted in the use of the hollow block floor slab.

This type of floor is based on the flanged beam principle and consists of concrete ribs 75 to 150 mm wide and suitably reinforced. Clay or concrete hollow

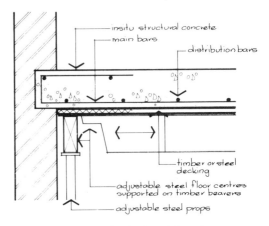

Figure 103 Suspended *in situ* reinforced concrete floor and formwork

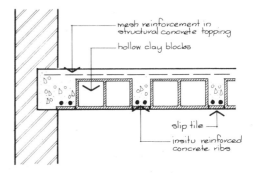

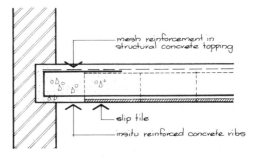

Figure 104 Suspended hollow block floor slab

blocks are placed between the ribs and an *in situ* concrete topping must be provided, see Figure 104.

It should be noted that the hollow blocks contribute very little to the strength of the floor, but they do assist in providing a level soffit for plastering. In some cases a slip tile is provided under the rib to reduce the problem of pattern staining.

14.4 Precast concrete floors

Precast concrete floor units can be divided into five groups as follows: Hollow units; solid units; trough or T section; beam-and-block and concrete planks used as permanent shuttering. These units are illustrated in Figures 105, 106 and 107.

The first three types require a crane for handling, but no props are required during assembly. A small amount of grout or concrete between the units is necessary and a structural concrete topping can be added to increase strength. A considerable advantage of these types of units is that they are able to carry their design load immediately after they have been erected.

The beam-and-block and concrete plank units require propping during construction and an *in situ* concrete topping is essential.

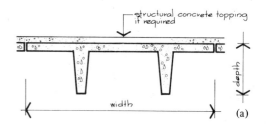

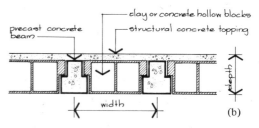

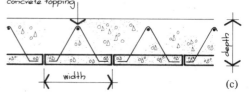

Figure 106   Precast concrete floor units: (a) tee section; (b) beam and block; (c) plank shuttering

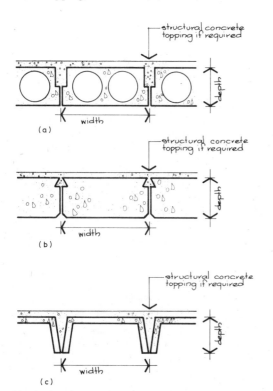

Figure 105   Precast concrete floor units: (a) hollow units; (b) solid units; (c) trough section

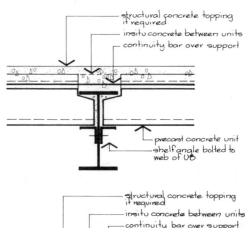

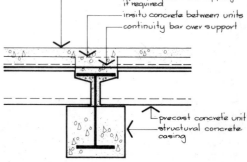

Figure 107   Details of supporting precast units on universal beam

Floor units are an economic form of construction for a very wide range of spans. By using double-T sections it is possible to obtain spans in excess of 13.0 m. In the short span range precast concrete planks with an *in situ* concrete topping provide an economic decking. There is a very wide range of floor units that can be used in the intermediate span range.

The precast concrete floor is not suitable for irregular plan shapes which would require a large number of different sized units.

Some types of units are used in conjunction with a suspended ceiling in which case services can be placed within the voids. In other cases it may be necessary to accommodate the services within the screed or structural topping.

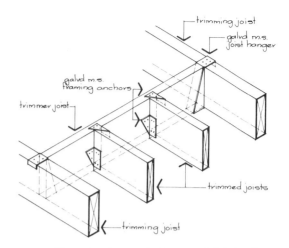

Figure 108 Trimming opening in suspended timber floor

14.5 Trimming openings

Figure 108 illustrates a method of trimming an opening in a timber floor. Various methods, such as a tusk tenon or a housing joint are used to connect the members together, but these have been generally superseded by the use of galvanised mild steel connectors which require less labour in forming the joint.

Openings in precast concrete floors can be formed either by providing trimming beams or by using a mild steel strap. The method to be adopted depends on the type of floor used. Figure 109 shows a method of trimming an opening in a beam and block floor.

Openings in concrete floors can be trimmed by increasing the reinforcement in the thickness of the slab around the opening, but in some cases it may be necessary to provide trimming beams if the opening is very large.

15. CONSTRUCTION OF A SIMPLE R.C. STAIR

15.1 Reinforced concrete stairs

Concrete stairs are widely used because of their high degree of fire resistance and the many variations of plan layout that are possible.

In situ concrete stairs may be constructed either by: canterlevering the flights from the supporting wall, spanning the flights between landings or beams, or the flight and landings acting as a single structural slab spanning between supports.

Where a crane is available, there may be some advantage in using precast concrete flights which span between *in situ* or precast concrete landings.

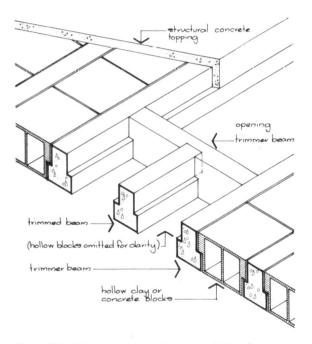

Figure 109 Trimming opening in beam and block floor

A wide variety of finishes can be applied to the treads and landings. The type will depend on the use of the stairs, but it should be noted that as the thickness of finish is greater on the floor than the treads, care must be taken when setting out the stair that the rise must be equal throughout.

Details of a simple reinforced concrete stair and landing are shown in Figure 110.

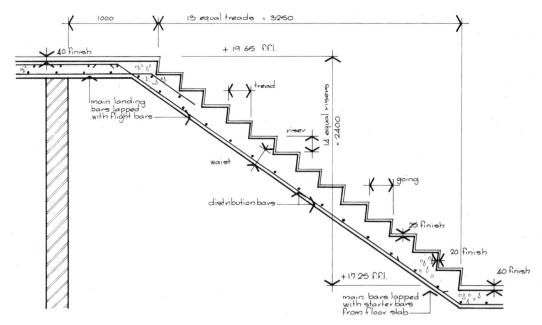

Figure 110   Simple reinforced concrete stair detail

Further reading

Readers who require a more in depth knowledge of some of the objectives in Topic Area D may find the following reading list useful.

D.14 McKay, J. K., *Building construction, Vol 3,* Longman Group Ltd (1974).

Boughton, Brian, *Reinforced concrete detailer's manual,* Crosby Lockwood Staples (1975).

D.15 Chudley, R, *Construction technology, Vol 2,* Longman Group Ltd (1974).

Boughton, Brian, *Reinforced concrete detailer's manual,* Crosby Lockwood Staples (1975).

Topic Area E – External Works

16. SURFACE WATER, SOIL DRAINS AND SEWERS UP TO 225 mm DIA

16.1 Pipe laying and testing

The work of constructing the sewer may begin when the line, levels and gradient of the sewer have been established, the position of the manholes has been decided upon and the materials specification has been finished.

Before the pipes can be laid to a true line and level it will be necessary for the trench to be excavated. To do this accurately with a bottom accurate to the falls shown on the drawings it is essential to set up sight rails to the line and gradients shown on the drawing. Ideally these should be painted black and white. A traveller should then be prepared and cut to the depth of the pipe invert plus the depth of bedding of the pipe and to this length must be added the height that the sight rails have been set above ground level, see Figure 11.1.

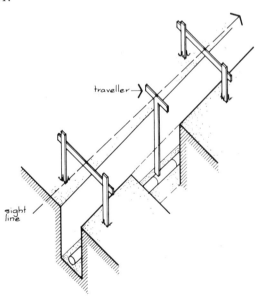

Figure 111 Typical arrangement in the use of sight lines and traveller

During excavation care should be taken to ensure that the excavated material is placed away from the edge of the trench to reduce the dead load on the vertical sides of the excavation, and avoid possible collapse. If there is any doubt at all that the trench sides may not be self supporting then they should be timbered to avoid danger to operatives working in the trench or along the line of excavation. In any event a trench deeper than 1.2 m in soil should be timbered.

Having completed excavation between manholes the pipe laying can begin. The pipes may be laid on a concrete bed or on a granular material depending on the specification.

Pipes with rigid joints are laid on a concrete bed which is either scooped out to receive the collars or tamped true to the falls, so that the pipes are supported on the collars. The pipes are then either carefully haunched or surrounded with concrete.

Pipes with flexible joints are usually laid on a granular bed and surrounded with a granular material (see Figure 112) which should be well compacted on

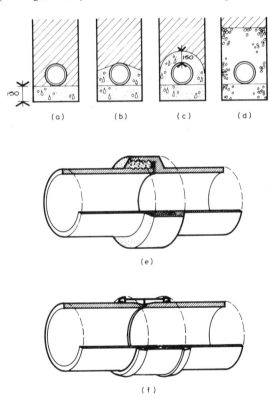

Figure 112 Pipe bedding and jointing: (a) concrete bed; (b) concrete bed and haunch; (c) concrete bed and surround; (d) granular bed and surround; (e) rigid joint; (f) flexible joint

either side of the pipe. Pipes are laid to a true line by drawing a taut string line either along the top of the pipe line, or to one side of the pipes. Whichever method is employed the pipes must be laid to a true line and gradient beginning at the lower manhole level and working towards the head of the sewer.

With rigid pipes, the spigot end of the pipe is first wrapped with gaskin or hemp and then placed into the socket of the pipe previously laid. Then by the use of a chalking tool and hammer, the gaskin is chalked into the joint of the pipe. It is then checked for line and level before sealing with a mortar of 1 : 2 cement/sand and finishing in a fillet around the outside of the pipe. In some cases it may be found more convenient to lay a number of pipes before sealing the joints with mortar.

On flexibly-jointed pipe laying, techniques are speeded by the use of 'push fit' type joints which do not require chalking or sealing with cement mortar. Because the pipes are supplied in longer lengths the number of joints can be reduced. These flexible joints are available with a number of different materials such as vitrified clay, concrete, asbestos cement, p.v.c. and pitchfibre. The type of joint and methods of jointing vary from one material to another and information is provided by the manufacturers on the correct method of jointing them.

On completion of the laying operation and before surrounding the pipeline and backfilling the trench the sewer line should be tested, to ensure that it is capable of carrying out the work for which it has been designed. A number of testing methods exist, the choice of test depending on the nature of the liquid to be carried by the pipeline.

(i) Water test

This is the most logical test since the sewer will eventually carry liquid. In this test a stopper or bag is inserted in the outfall end of the pipeline and the line filled with water to provide a head of water of between 1.2 and 6.0 m.

An appropriate period should be allowed for absorption of water by the pipes, then the loss of water over a period of 30 min, should be measured by adding water from a measuring vessel, at regular intervals and noting the quantity required to maintain the original head. CP 2005: 1968 sets out the criteria for the test.

(ii) Air test

This type is considered a more severe test than the previous method, but it is a more convenient way of testing drains and sewers before backfilling.

CP 2005: 1968 requires that the length of pipeline under test should be effectively plugged and air pumped in by a suitable means, e.g. a bellows or air pump, until a pressure of 100 mm in a glass U-tube connected to the pipeline is indicated. The air pressure should not fall to less than 75 mm during a period of 5 minutes.

(iii) Smoke tests

A smoke cartridge is introduced into the line of pipes to be tested and then both ends of the line are plugged. An air line is then used to increase the pressure in the drain.

As an alternative, a smoke-making machine may be used in which cotton waste or similar material is burned in a chamber to produce smoke. This is pumped into the line of pipes by a bellows which forms an integral part of the machine.

(iv) Miscellaneous tests

Other tests may include a *light test* which employs the use of a light at one end and a mirror tilted at 45° to the line of pipe at the other manhole. If the pipeline is free from obstruction and laid true a clear image of the light should appear in the mirror.

A *steel ball* is sometimes used to run down a line of pipe to check that the line is free from obstruction, the ball is 13 mm smaller than the diameter of the pipe.

On existing sewers where it is thought there may be an obstruction or other problem a *television survey* may be carried out.

16.2 Road gullies

Rain falling on the carriageway and footways of the road network is collected into the channels and discharged through gullies into the sewer system.

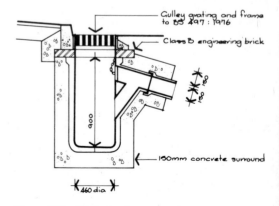

Figure 113 Typical gulley arrangement

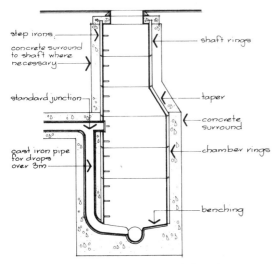

Figure 115(a)   Precast concrete manhole and backdrop

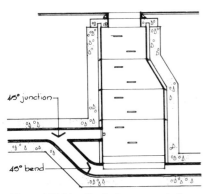

Figure 115(b)   Typical ramp manhole up to 1.5 m drop

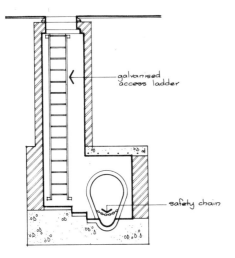

Figure 115(c)   Side entrance manhole on large sewer

Gullies are usually spaced so that their maximum catchment area is approximately 190 m². The exact spacing depends on the width of road and footways and whether the road is cambered or is a straight crossfall. In a section of road with a straight crossfall gullies are only required on the lower channel.

Gulley gratings and frames are manufactured to BS 497: 1976 and may be sited either in the channel or as side entry gullies with the gully face conforming to the profile of the kerb, see Figure 113.

Gully pots should conform to either BS 556, Pt. 2: 1972 (Concrete) or BS 539: 1971 (Salt glazed), they may also be manufactured from u.p.v.c., for which there is no British Standard at present, or be built *in situ* of brickwork. No attempt should be made to cut costs by omitting gullies and where

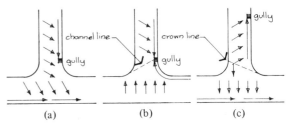

Figure 114   Drainage at junctions: (a) crossfall in one direction; (b) crossfall resulting in channel line; (c) crossfall resulting in crown

gradients are steeper than 5% or flatter than 0.5% additional gullies should be provided. The gullies should be sited just uphill of the tangent point at road junctions so that surface water in the channel does not flow across the junction, see Figure 114.

16.3 Manholes
Manholes can be divided into two classes, simple manholes for use on small sewers, and the special manholes on large and usually deep sewers. These include backdrop manholes, side entrance manholes, storm overflows and pressure manholes. This class of

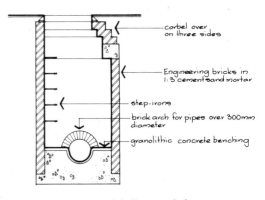

Figure 115(d)   Typical shallow manhole

manhole construction is expensive to build and was in the past almost invariably built in brickwork.

Mass concrete, precast concrete rings and reinforced concrete are now the materials which are in common use and cheaper to construct than traditional brickwork. A combination of these materials is often used. Figure 115 illustrates some of the types of manhole in use.

16.4 Types of manhole

(i) Brickwork manholes

These are usually constructed as a rectangular chamber on a concrete foundation with either a reinforced concrete roof slab or a brick arch, and an access shaft 675 × 750 mm in size. Class B engineering bricks to BS 3921: 1974 laid in English bond on a 1 : 3 cement/sand mortar bed and pointed flush are used to construct the manhole.

In good ground 225 mm brickwork is adequate for depths up to 3 m but an increase in wall thickness will be necessary, for extra depth, or unusual ground conditions, or to resist water pressure. Step irons are built into the brickwork on a 300 mm staggered line. But for depths greater than 5 m galvanised iron access ladders are provided.

(ii) Precast concrete ring manholes to BS 556: 1972

These may be built on a precast or *in situ* concrete base. They can be constructed rapidly using unskilled manpower. This type is particularly effective in poor ground conditions and is surrounded with 150 mm of concrete to ensure watertight conditions. Taper sections are available to reduce the chamber size to a 675 mm diameter shaft.

Step irons are provided in the precast concrete ring sections, but manholes deeper than 5 m are provided with access ladders of galvanised iron.

Manholes are provided at all changes of direction, gradients, and pipe size, at junctions and at heads of all sewers. The maximum distance apart of manholes for small sewers should be 100 m although this distance may be increased for sewers over 1 m diameter.

16.5 Connection to sewer
The Public Health Act 1936 and the Building Regulations lay down the legislation for connecting drains to Local Authority sewers. The methods of connection depend upon the position of the drain in relation to the sewer, the size and condition of the existing sewer, the material the sewer is constructed from and the difference in invert levels.

Whichever method is chosen the drain entering the sewer must discharge its contents obliquely in the direction of flow of the sewer. The connection must be made so that it will remain watertight under all flow conditions and will work satisfactorily.

The methods of connection may be at an existing or new manhole where the gradient of the contributing drain can be increased to allow the connection to be made at invert level, or by the use of a ramp if the invert level difference is less than 1.5 m. If it is not economical to increase the gradient of the new drain and a ramp will not fulfil the function a backdrop manhole may be constructed.

In the event of a new sewer being constructed to serve an area, 'stopper' junctions are provided in the line of the sewer to allow connections to be made from existing properties.

If a new property were to be built that had not been allowed for, then on a small sewer up to 225 mm dia serving an area, an oblique junction could be inserted in the line by removing two or three pipes and replacing one of the straight pipes with the junction, then connecting the new property to this

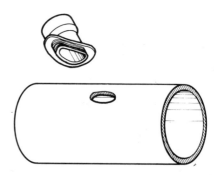

Figure 116(a)   Saddle connection

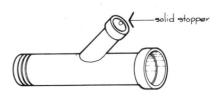

Figure 116(b)   Stopped junction

junction. To connect a property to a larger sewer a saddle connection (see Figure 116) is made by carefully breaking a hole in the top half of the sewer pipe and trimming the opening so that the saddle pipe fits neatly onto the sewer pipe. It is then jointed up with cement mortar and surrounded with concrete.

16.6 Safety precautions

The dangers to operatives working in sewers is greater than is generally realised. The risks to which the sewer worker is exposed can be summarised as follows:

(i)   Falling from ladders or step irons.
(ii) Injury from tools being dropped down a manhole.
(iii) Injury by falling while travelling in a sewer.
(iv) Being swept away or drowned.
(v)   Infection (Weils disease). Leptospiral Jaundice.
(vi) Poisoning by gasses or vapour.
(vii) Physical injury by explosion.

The way to minimise accidents in sewers is to train operatives in safe working practice, to recognise potential hazard and, in the case of an accident, the procedure to be followed. The safety precautions to be exercised in sewers are beyond the scope of this topic area but may be found in the Local Government Training Board's book *Safety in sewers* and further reference may be made to the Institute of Civil Engineers booklet *Safety in sewers and sewage works.*

17. CALCULATION OF SURFACE WATER RUN-OFF AND DISCHARGE

17.1 Surface water

The provision of adequate surface water drainage facilities are an essential part of any road construction scheme. These must cater for the run off from the carriageway, footway, verges and hard shoulders on motorways, and for the run off from catchment areas adjacent to the road line, which may be affected by the construction.

Road Note 35 (1976) *A guide for engineers to the design of storm sewer systems* sets out recommended guidelines for the design of drainage systems for small areas such as housing estates where the largest sewer will probably not exceed 600 mm dia and the Transport and Road Research Laboratory Hydrograph method which will provide accurate sewer design for urban areas.

17.2 Surface water calculations

Before surface water calculations may be made a number of terms on which these calculations are based should be understood.

(i) Catchment area. This is the total area from which run off of all surface water would flow by gravity to a sewer.

(ii) Time of concentration. This is the time taken for water to reach the point under consideration after falling on the most remote part of the surface. Its value is given by the sum of the time taken to flow across the surface and enter the sewer (time of entry) and the time to flow along the sewer assuming full bore velocity.

(iii) Time of entry. It is recommended that a time of entry of two minutes should be used for normal urban areas increasing up to four minutes for areas with large paved surfaces and slack gradients.

(iv) Time of flow along the sewer is calculated assuming full bore velocity.

(v) Intensity of rainfall. This depends on storm duration and frequency of storm and it is appropriate to use a mean rate of rainfall during a storm. The duration of the storm should be taken as being equal to the time of concentration of the drainage area, to the point for which the calculation is being made. In the absence of precise local data the Ministry of Health formula may be used for most road drainage systems

$$R = \frac{750}{t + 10} \text{ mm/hour for storms of 5 to 20 min duration.}$$

$$R = \frac{1000}{t + 20} \text{ mm/hour for storms over 20 min duration.}$$

where t = time of concentration

It is suggested in Road Note 35 that the following is adopted when using the Rational (Lloyd Davies) formula for the calculation of surface water sewers up to 600 mm diameter.

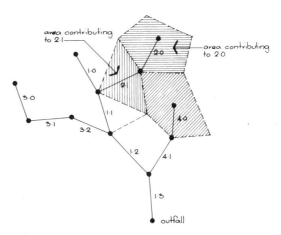

Figure 117   Example of decimal classification (*From Road Note 35*)

RATIONAL FORMULA DESIGN SHEET
Storm frequency one in 1 year
Time of entry 2 min/roughness coefficient 0.6 mm

Example of pipe length 1.1

CRIMP AND BRUGES VALUE (   )

1	2	3	4	5	6	7	8	Impermeable area (ha)				13	14	15
Pipe length (number)	Diff. in level (m)	Length (m)	Gradient (1 in)	Velocity (m/s)	Time of flow (min)	Time of concentration	Rate of rainfall (mm/h)	9 Roads	10 Bldgs. yards etc.	11 Total (9 + 10)	12 Cumulative	Rate of flow (l/s)	Pipe dia. (mm)	Remarks
1.0	1.10	63.1	57	(1.25) 1.33	0.79	2.79	67.9	0.089	0.053	0.142	0.142	(22.8) 26.8	150	
1.1	1.12	66.1	59	(1.61) 1.70	0.65	3.44	62.5	0.077	0.109	0.186	0.328	(66.2) 56.9	225	
1.2	0.73	84.7	116	(1.40) 1.46	0.97	4.41	57.4	0.081	0	0.081	0.409	( )		

Length number 1.1 Assume pipe size 150 mm dia   Column 1
Length 66.1 – Difference in level 1.12 m   Columns 2 and 3
Gradient $L \div F$ 66.1 ÷ 1.12 = 1 in 59   Column 4
Velocity of flow from tables 1.70 (1.61 Crimp & Bruges)   Column 5
Time of flow 66.1 ÷ 1.70 ÷ 60   = 0.65   Column 6
Add time of entry   = 2.00   Column 6
Add time of flow of 1.0   = 0.79

Total time of concentration so far = 3.44 min   Column 7

$$Q = \frac{Ap \times 1}{0.360} = \text{litres/sec}$$

$$Q = \frac{.328 \times 62.5}{0.360} = 56.94 \text{ litres/sec}$$

Figure 118 Design sheet with worked example (*From Road Note 35*)

(i) A key plan of the proposed sewer system should first be prepared (see Figure 117) using a decimal classification. In this way, if it becomes necessary to add a further branch it does not matter as far as the calculation is concerned.

(ii) Prepare a design sheet to simplify calculations. Refer to Figure 118.

(iii) Enter the basic design data on the design sheet, columns 1, 2, 3, 4, 9, 10 and 11, the total in column 11 is the area of surface directly connected into each length and is given by the sum of columns 9 and 10. Column 12 can now be completed by adding in the contributions from the branches at the appropriate junctions. This column gives the total area of surface contributing to the flow in a length. Reference to Road Note 35 will clarify the method of entry of the branches of the sewer onto the design sheet.

(iv) A pipe size (column 14) is now assumed, the pipe full velocity of flow (column 5) is found from published tables* and the time of low (column 6) calculated from columns 3 and 5. The time of concentration (column 7) is the total time of flow up to and including the length under consideration plus the time of entry say 2 min, where sewers join the time of concentration is taken to be the greatest time to the manhole concerned.

(v) The rate of rainfall (column 8) corresponding to the time of concentration is found from tables published by the Meterological Office or an approximation may be obtained from the Ministry of Health formula.

(vi) The expected peak rate of flow in the pipe is then given by the 'rational' formula.

$$Q = \frac{Ap \times i}{360} \; \text{m}^3/\text{sec} \quad \text{or} \quad Q = \frac{Ap \times i}{0.360} \; \text{litres/sec}$$

where

Q = run-off litres/sec or m^3/sec;
Ap = impermeable area in hectares;
i = intensity of rainfall mm/hour

using the figures given in columns 8 and 12.

The assumed pipe size is then checked to see if it can carry the expected flows, if not a larger pipe is assumed, and the steps from (iv) are carried out again until a pipe of sufficient size is reached.

*1. Tables for the hydraulic design of storm drains, sewers and pipe lines, 2nd Edition.
2. Crimp and Bruges, Tables and diagrams for use in designing sewers and water mains.

18. FLEXIBLE AND RIGID PAVEMENT AND METHODS OF SURFACING

18.1 Flexible and rigid construction

Flexible pavements are a development of the Telford or Macadam principle of construction. They are of a layered construction which relies on the interlock of the particles in each layer for stability, and on the load spreading properties of the layers to distribute the wheel loadings to the subgrade.

Rigid pavements are of concrete slab construction, usually resting on a layer of granular material. They distribute the wheel loadings over a wider area and are often used on the weaker subgrades.

The choice of construction is influenced by a number of factors which are taken into account at the design stage. These include the present traffic flow, and the expected growth rate over a period of up to forty years, the bearing capacity of the subgrade, the level of the water table below the formation, and more recently, since the escalation of the price of oil, from which bitumen is a by-product, the economics of construction and maintenance. A few years ago about 90% of the new road construction in this country was of the flexible type, but since the oil crisis more construction of the rigid type of pavement has taken place.

Road Note 29 (3rd ed) 1970 *A guide to the structural design of pavements for new roads*, gives examples of both flexible and rigid design and sets out tables and graphs for the design of pavements from residential roads to motorways. From initial traffic intensities and estimated growth rate designs can be prepared suitable for any life up to forty years.

18.2 Explanation of terms

A glossary of highway engineering terms can be found in BS 892: 1954. A number of the terms directly related to flexible and rigid construction in Figure 119 are described below.

Road pavement. The total depth of construction resting on the subgrade which will support the traffic loads.

Subgrade. The natural foundation or fill which receives the loads from the pavement.

Formation. The prepared surface of the subgrade on which the pavement is constructed.

Sub-base. A secondary layer of material provided between the formation and the base, or concrete slab.

Road base. The layer which provides the principal support for the surfacing. In rigid construction this layer is called the slab, if no surfacing layer is provided.

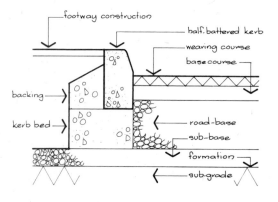

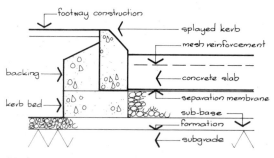

Figure 119 (top) flexible construction; (bottom) rigid construction

Surfacing. The top layer or layers of the pavement comprising:

(a) *Wearing course.* The top layer of the surfacing which carries the traffic.

(b) *Base course.* The layer sandwiched between the wearing course and the road base. It is not always necessary to include a base course since surfacing may be laid as a single course.

Any reference to other terms which are not clear to the reader may be found in BS 892: 1954.

18.3 Design criteria

Although a number of design methods are in use to determine the thickness or a road pavement, current UK practice is to design in accordance with the methods laid out in Road Note 29 for both flexible and rigid construction. There are two basic criteria on which design of the pavement is based, these are:

1. The quality of the subgrade;
2. The cumulative number of standard axles to be carried during the design life of the road in the single lane of a two lane carriageway, or the slow lane of a dual carriageway.

The quality of the subgrade is the principal factor in determining the thickness of the pavement. Deterioration by frost action must also be taken into account at the design stage, since this could have serious consequences on the bearing capacity of the subgrade. The strength of the subgrade is assessed on the California Bearing Ratio (CBR value, a test developed in the USA). The test is described in full in BS 1377: 1975 *Methods of testing soils for civil engineering purposes.*

The testing apparatus is of the type shown in Figure 120. The samples of soil must be compacted to a dry density similar to that which it is expected to achieve in practice on site. The sample having been prepared to simulate the maximum moisture content likely to be experienced at the completion of the road works in the subgrade. The test results are

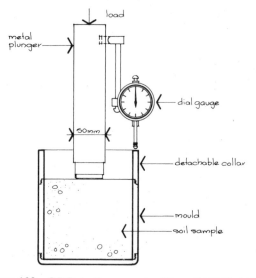

Figure 120 C.B.R. testing apparatus (*From BS 1377: 1975*)

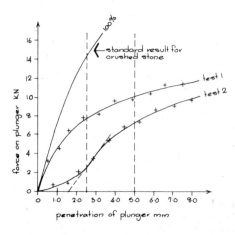

Figure 121 Typical C.B.R. test results (*From BS 1377: 1975*)

plotted on a load penetration graph, Figure 121 and a line is drawn through the points plotted. The curve is usually smooth towards the zero point, but occasionally a correction has to be made (test 2).

If the line of curve from the zero mark is concave, this entails drawing a tangent to the curve at the steepest point of the curve, and projecting the line down to the penetration axis. At the point where this line cuts the axis a new point of origin is placed for the test. The loads required to produce penetrations at 2.5 mm and 5 mm are recorded and expressed as ratios of the loads required to cause the same penetrations in a standard crushed rock material.

18.4 Sub-grades

Road Note 29 recognised that roads will have to be built over relatively weak subgrades. A table which correlates between CBR values and soil types based

Table 12   Table of CBR values for soil found in UK

Type of soil	Plasticity index (per cent)	CBR (per cent)	
		Depth of water-table below	
		More than 600 mm	600 mm or less
Heavy clay	70	2	1*
	60	2	1.5*
	50	2.5	2
	40	3	2
Silty clay	30	5	3
Sandy clay	20	6	4
	10	7	5
Silt	—	2	1*
Sand (poorly graded)	non-plastic	20	10
Sand (well graded)	non-plastic	40	15
Well-graded sandy gravel	non-plastic	60	20

* See para. 27 Road Note 29.
(*From Road Note 29*)

on British experience with a wide variety of subgrades and moisture conditions for high and low water tables is shown in Table 12. The soils for pavement design are classified into seven main-types (Table 12) and these fall into two categories:

1. Cohesive soils (plastic) which are clays and mixtures of clay and sand or silt.
2. Non-cohesive (non plastic) soils which are sands and gravels, coarsely grained materials.

Coarsely grained soils are not generally frost susceptible, since the voids of the compacted material are sufficient to take up any expansion of soil and water due to freezing. Not all cohesive soils are susceptible to frost action, those which have fine grains (particles finer than 0.002 mm) are impermeable and therefore not susceptible. Frost susceptible materials usually have at least 10% to 15% of particles by weight smaller than 0.02 mm silts, fine sands and some chalks. To reduce the risk of frost damage to sub-grades, it may be necessary to increase the constructional depth of the pavement by at least 450 mm above formation.

18.5 Compaction of sub-grade

Roads tend to be built to maximum gradients of approximately 1 : 25 (4%). This may involve a certain amount of balancing of the earthworks between cut and fill as well as founding the road at existing ground levels. In this process soils are excavated at one point of the road line, where there is a surplus of material and transported to, spread and compacted at another point where the original profile of the ground is lower.

Compaction of the soil reduces the air voids in the soil, and so reduces the risk of moisture change in the subgrade. If the compaction is not carried out correctly during the earthworks stage of the contract, then traffic using the road at a later date will cause further compaction to take place which will distort the road pavement.

Table 13 which is taken from the Department of Transport Specification for Road and Bridge works lists the type of compaction plant, its category and the minimum number of passes for the compacted depth of soil type. Comparative field density tests carried out in accordance with BS 1377: 1975 (test No. 15) on material which is suspected of not being compacted adequately, and compared with tests made on adjacent approved work may show that further compaction is required in the area.

Another factor which may affect the strength of the subgrade is the level of the water table. An increase in moisture content can reduce the strength and bearing capacity of the subgrade.

Road Note 29 distinguishes between a water table within 600 mm of the formation and a water table more than 600 mm below formation. Ideally the water table should be a minimum of 1.2 m below formation, but moisture movement due to seasonal changes can take place in a number of ways, see Figure 122.

To reduce such changes and maintain a stable moisture content in the subgrade, where the need

Table 13 Compaction plant and maximum depth of layers (*From Specification for Roads and Bridge Works*)

Compaction requirements — D = Maximum depth of compacted layer (mm); N = Minimum number of passes

Type of compaction plant	Category	Cohesive soils		Well graded granular and dry cohesive soils		Uniformly graded material	
		D	N	D	N	D	N
Smooth wheeled roller	Mass per metre width of roll:						
	Over 2100; up to 2700 kg	125	6	125	10	125	10*
	Over 2700; up to 5400 kg	125	6	125	8	125	8*
	Over 5400 kg	150	4	150	8	Unsuitable	
Grid roller	Over 2700; up to 5400 kg	150	10	Unsuitable		150	10
	Over 5400; up to 8000 kg	150	8	125	12	Unsuitable	
	Over 8000 kg	150	4	150	12	Unsuitable	
Tamping roller	Over 400 kg	225	4	150	12	250	4
Pneumatic-tyred roller	Mass per wheel:						
	Over 1000; up to 1500 kg	125	6	Unsuitable		150	10*
	Over 1500; up to 2000 kg	150	5	Unsuitable		Unsuitable	
	Over 2000; up to 2500 kg	175	4	125	12	Unsuitable	
	Over 2500; up to 4000 kg	225	4	125	10	Unsuitable	
	Over 4000; up to 6000 kg	300	4	125	10	Unsuitable	
	Over 6000; up to 8000 kg	350	4	150	8	Unsuitable	
	Over 8000; up to 12000 kg	400	4	150	8	Unsuitable	
	Over 12000 kg	450	4	175	6	Unsuitable	
Vibrating roller	Mass per metre width of a vibrating roll:						
	Over 270; up to 450 kg	Unsuitable		75	16	150	16
	Over 450; up to 700 kg	Unsuitable		75	12	150	12
	Over 700; up to 1300 kg	100	12	125	12	150	6
	Over 1300; up to 1800 kg	125	8	150	8	200	10*
	Over 1800; up to 2300 kg	150	4	150	4	225	12*
	Over 2300; up to 2900 kg	175	4	175	4	250	10*
	Over 2900; up to 3600 kg	200	4	200	4	275	8*
	Over 3600; up to 4300 kg	225	4	225	4	300	8*
	Over 4300; up to 5000 kg	250	4	250	4	300	6*
	Over 5000 kg	275	4	275	4	300	4*
Vibrating-plate compactor	Mass per unit area of base plate:						
	Over 880; up to 1100 kg	Unsuitable		Unsuitable		75	6
	Over 1100; up to 1200 kg	Unsuitable		75	10	100	6
	Over 1200; up to 1400 kg	Unsuitable		75	6	150	6
	Over 1400; up to 1800 kg	100	6	125	6	150	4
	Over 1800; up to 2100 kg	150	6	150	5	200	4
	Over 2100 kg	200	6	200	5	250	4
Vibro-tamper	Mass:						
	Over 50; up to 65 kg	100	3	100	3	150	3
	Over 65; up to 75 kg	125	3	125	3	200	3
	Over 75 kg	200	3	150	3	225	3
Power rammer	Mass:						
	100; up to 500 kg	150	4	150	6	Unsuitable	
	Over 500 kg	275	8	275	12	Unsuitable	
Dropping-weight compactor	Mass of rammer over 500 kg Height of drop:						
	Over 1; up to 2 m	600	4	600	8	450	8
	Over 2 m	600	2	600	4	Unsuitable	

For items marked * the roller shall be towed by a track laying vehicle.

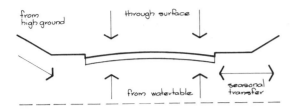

Figure 122 Ways in which water can enter and leave the subgrade

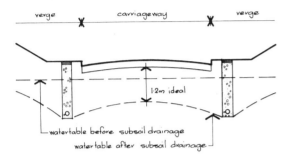

Figure 123 Effect of subsoil drainage on water table

arises, sub-soil drains are provided parallel to the pavement, and on either side as shown in Figure 123.

Preparation and surface treatment of the formation, should only be carried out after completion of the subgrade drainage, and immediately prior to laying the sub-base or road base material. This treatment should take the form of reinstatement of any soft areas in the subgrade, and the removal of mud and slurry. The surface should be compacted with a smooth wheeled roller, and the formation trimmed to shape and rolled with one pass of a smooth wheeled roller.

Where the formation is not to be protected immediately by the sub-base or road base, it should be covered with a plastic sheeting to prohibit the ingress of water.

To protect the formation before final shaping it is usual to remove material to the full width of the carriageway, and to within 300 mm of formation level. The final trimming to formation level is carried out in one operation and construction traffic should be prohibited on the formation.

18.6 Materials

When the subgrade has been prepared the sub-base should be constructed as quickly as possible and only essential traffic should be allowed on the formation prior to the pavement being constructed.

The function of the sub-base is as a structural layer which will bear greater stresses than the subgrade, and,

in the case of a subgrade which may be frost susceptible it has a role of acting as an insulating layer as well as providing a temporary road which the construction vehicles may use.

Sub-base material is usually brought on to site by side or end tipping lorries which discharge their loads onto the subgrade. The material is then spread using a motor grader to give a uniform thickness, and rolled in the manner specified in Figure 124, except for lean concrete and dry bround macadam. Where the sub-base or road base are unbound materials, the top

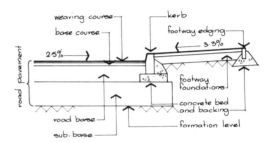

Figure 124 Typical flexible construction detail

75 mm should be scarified, re-shaped and re-compacted to obtain compliance with the specification for profile and tolerance.

The Department of Transport Specification for Road and Bridge Works defines two unbound granular sub-bases both of which when compacted give adequate strength requirements for pavement design.

Type 1 materials for sub-bases are crushed concrete, crushed slag, crushed rock or well burnt non plastic shales.

Type 2 materials for sub-bases are natural sands and gravels.

Where type 1 and type 2 material is not readily available, cement-bound materials may be used if it is economical to do so. They are divided into soil cement, cement-bound granular material and lean concrete. Since these may also be used as a road-base material they will be described under this heading.

The road-base is the principal load carrying layer, which supports the surfacing and reduces stresses on the sub-base and subgrade. The Department of Transport specifies three groups of materials.

1. Unbound material.
2. Cement bound materials.
3. Bituminous materials.

1. Unbound material

Unbound bases are specified as:

(i) Wet-mix macadam a crushed rock or slag material of specified grading with a sufficient moisture content to give maximum compaction as determined by tests carried out to BS 1377. The material is spread in layers not exceeding 200 mm thick and compacted in accordance with Table 13.

(ii) Dry bound macadam is a layer of single size crushed slag or rock of 50 or 40 mm nominal size which is uniformly laid to a thickness of between 75 and 100 mm and given two passes of a smooth wheeled roller. Fine aggregate is then spread on it to a thickness of 25 mm and vibrated into the voids by a vibrating roller or vibrating plate compactor. This operation is repeated until no more will penetrate into the surface. It is then brushed to remove excess fine material and the whole operation repeated until the full specified thickness is reached.

(iii) Crusher run road bases depend for stability on the particle interlock of the material, these materials cannot be closely controlled for grading and an attempt is made to ensure the presence of particles of various sizes from 75 mm maximum size down. It is brought to site and compacted in layers of 100–150 mm.

Where this material is readily available in some districts, and therefore relatively cheap, it is specified by the authorities. Because of the low moisture content when the material is brought to site maximum compaction cannot be achieved and some deformation under traffic conditions after the road is opened may be experienced.

(iv) Colliery shale road bases have been used for lightly trafficked roads. The material must be well burnt, not likely to soften in water and not be frost susceptible. Accumulation of these waste materials often leads to them being available at low cost and there is sometimes pressure for them to be used in road construction but it must be borne in mind that they should only be used if it is certain that they will fulfil their function satisfactorily and not be affected by moisture or frost action.

(v) Hardcore if used in road construction is usually specified as clean rock-like material, such as broken concrete and sound broken brick free from mortar, plaster and wood. Hardcore has been used on factory roads and estate roads, it is laid on a granular material which it can bed into and it needs heavy rolling, a minimum thickness of 150 mm should be specified and any 'hungry' areas are blinded with a fine hard material to ensure a closed surface finish.

2. Cement bound bases

These are specified as:

(i) Soil cement. This process requires the complete mixing of the soil with cement to give an average crushing strength of 2.8 MN/m^2 after seven days curing on a 150 mm cube.

Basically the cement is used as a binder to strengthen the soil, to do this a more careful control must be exercised over the specification and works than on other cement bound bases. The 'mix in place' method involves:

(a) Pulverising the soil to a depth of 200 mm.
(b) Spreading a uniform layer of cement to give the required strength after compaction.
(c) Adding water as necessary to meet the compaction requirements.
(d) Mixing together the soil-cement and water to the full depth.
(e) Compacting with a suitable roller.

A large variety of plant is available for this process from agricultural machines to purpose built plant. As an alternative to the 'mix in place' method the soil may be taken to a 'stationary plant' and mixed with the cement in paddle or pan type mixers, brought back to site, laid by a bituminous type paver and compacted.

(ii) Cement bound granular material is a granular material mixed with cement. The aggregate can be a naturally occurring gravel-sand, a washed or processed granular material, crushed rock or slag or any combination of these. It is sufficiently well graded to give a well closed surface finish after compaction in accordance with Table 13.

The cement content must be sufficient to give an average crushing strength based on 150 mm cube of 3.5 MN/m^2. This material like cement soil is mixed in pan or paddle mixers which are considered more efficient for this type of material.

(iii) Lean concrete can be produced in ordinary free fall concrete mixing plants as well as paddle type mixers. The aggregate to cement ratio is usually between 1 : 15 and 1 : 20 and the average twenty-eight day crushing strength of a group of three 150 mm cubes should be such that not more than one in any consecutive five such averages is less than 10 MN/m^2 or more than 20 MN/m^2.

The aggregate consists of either coarse and fine aggregate batched separately or an all-in aggregate with a maximum nominal size not exceeding 40 mm.

When the material has been mixed it is carried to the site in vehicles suitable for the task. During transport and while awaiting tipping it is protected from the weather by sheeting the vehicles. The material is usually laid by bituminous pavers adapted for the purpose but on many sites it is spread by grader or angle dozer, raked to profile and roller by a smooth wheeled roller to attain the required degree of compaction.

3. Bituminous road bases

The fifth edition of the Department of Transport Specification for Road and Bridge Works (1976) lists three types of coated material for use in sub-bases:

(i) Dense tarmacadam road base. This material is to comply with the general requirements of BS 4987: 1973. There are particular binder requirements to be observed when using flint gravel aggregates and a minimum layer thickness of 60 mm may be used to achieve the specified road base thickness.

(ii) Dense bituman macadam road base. This material must also comply with BS 4987: 1973, where gravel other than limestone is the aggregate the material passing the 75 μm sieve should include Portland cement or hydrated lime.

(iii) Rolled asphalt road base to BS 594: 1973. Rolled asphalts are designed in a different manner from coated macadams. The binder normally used for rolled asphalt road-base is petroleum bitumen of penetration 50, (50 pen), the mixing temperature of the dried aggregates is between 150°C and 205°C and the binder temperature at mixing is up to 175°C.

On discharge from the mixer the temperature should not exceed 190°C and at the site of laying the temperature should be between 125 and 190°C. All these temperatures are much higher than those for dense bitumen and tarmacadams and this reflects the harder binder used in the asphalt.

These coated materials should be laid by a bituminous paver in layers to give a maximum compacted thickness of 100 mm. Compaction is by means of an 8–10 tonne smooth wheel roller or by a pneumatic roller of equivalent weight. A feature of dense bituminous road bases is their load spreading qualities as compared to uncoated materials, a further advantage may be derived from an early application of the bituminous materials since this will protect the sub-base and subgrade from the ingress of water.

Bituminous surfacing materials

The functions of bituminous surfacings are:

(a) To prevent the ingress of water.
(b) Resist deformation due to traffic stress.
(c) Provide a highly skid resistant surface.
(d) Provide a satisfactory riding surface.

Surfacing can be laid as a single course but it is usually laid as two separate elements of basecourse and wearing course (Figure 124) the basecourse being about 60% of the total thickness of surfacing.

(i) Rolled asphalts to BS 594: 1973

Rolled asphalt is made by mixing a bitumen of low penetration with a graded crushed rock, slag, or gravel aggregate together with crushed or natural sand and a filler such as limestone dust or Portland cement.

Wearing course mixtures vary in coarse aggregate content from 30% to 35% depending on the intensity of traffic expected. Base course mixtures have a much higher coarse aggregate content of about 60% with a lower percentage of binder and little or no filler. Coated chippings of 14 or 20 mm nominal size are rolled into the surface as part of the laying operation to give a skid resistant surface.

(ii) Dense coated macadam to BS 4987: 1973

These materials can be made with tar or bitumen binder, they are almost impervious to water but may need to be surface dressed to give a satisfactory resistance to skidding. The aggregate for the wearing course is usually 10 or 14 mm and for both basecourse and wearing course the aggregate may be crushed rock, gravel or slag with a filler which is usually ground limestone dust.

The material is laid by a paver or by hand in areas such as footways or in regulating courses, but in these cases the binder viscosity is normally lower to allow the material to be handled and adequately compacted.

(iii) Open textured coated macadams to BS 4987: 1973

A feature of these types of material which are used for both base course and wearing course is that the material has a low fines content not exceeding 25%. This, while making them easy to lay, results in an open textured surface. The binders are in general less viscous than those used for dense macadams and this permits them to close up under the action of traffic, thereby becoming stronger and less permeable. In the main they are used as a surfacing material on lightly trafficked roads.

(iv) Cold asphalt surfacing

Fine and coarse cold asphalts are used to provide a wearing course on footways and lightly trafficked roads. The material is laid to a thickness not greater than 20–25 mm and compacted with a smooth wheeled roller. The name implies these materials are cold but in fact for surfacing purposes they are laid warm.

(v) Dense tar surfacing (DTS) to BS 5283: 1975

Dense tar surfacing is impervious to water and is not greatly affected by the spillage of diesel from vehicles. It is a hot process material consisting of a mixture of coarse and fine aggregate, filler and a high viscosity tar.

The material can be used for surfacing new roads with a medium traffic flow, and for resurfacing existing carriageways carrying normal commercial vehicles of up to 2000 per day (sum in both directions). The material should be laid by paver, but on small sites it can be laid by hand.

(vi) Pavement quality concrete

The concrete pavement slab thickness will depend on the expected traffic intensities over the design life of the road and to some extent on the bearing capacity of the sub-grade. As previously stated the methods of estimating this for design purposes can be found in Road Note 29.

Pavement quality concrete must meet specification requirements for cement content, strength, and air entrainment where required. The Department of Transport specification requires a cement content of ordinary portland cement or portland blast furnace cement of not less than 280 kg/m^3 of fully compacted concrete.

The relationship of the materials and their position in the road pavement is shown in Figure 125.

18.7 Methods of pavement construction

Flexible pavements rely on the load spreading properties of a layered system of construction to distribute the wheel loads, imposed on the pavement surface, over the sub-grade. A rigid pavement consists of a concrete slab, resting on a relatively thin sub-base, which acts like a plate to distribute the wheel loads over a much wider area of subgrade.

Because of these two essentially different methods of distributing the stresses applied to the pavement surface, over the subgrade, different methods of construction are employed.

Flexible construction

Two preliminary requirements to the satisfactory laying and compaction of coated materials are that:

(i) They should be delivered to site at a suitable temperature.
(ii) The surface on which they are to be laid should be structurally sound and of the correct profile.

With the exception of mastic asphalt all pre-mixed bituminous materials for pavement construction may be laid by machine although hand laying can be carried out on small sites, and in awkward areas. The material is delivered to site in lorries with insulated tipping bodies covered by tarpaulins. It should remain sheeted until required.

The bituminous material is fed from the lorry into a mechanical spreader and finisher (Figure 126) capable of laying to the required widths, profile, camber or crossfall without causing segregation or dragging of the material.

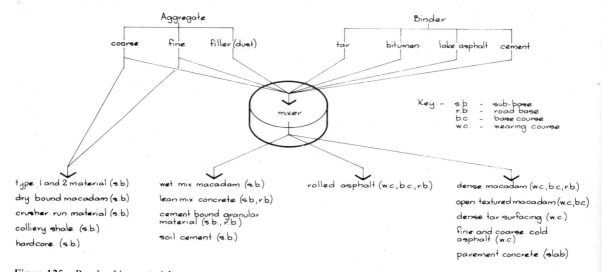

Figure 125 Roadmaking materials

84

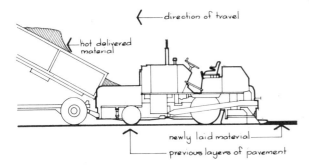

Figure 126 'Flexible Paving' machine

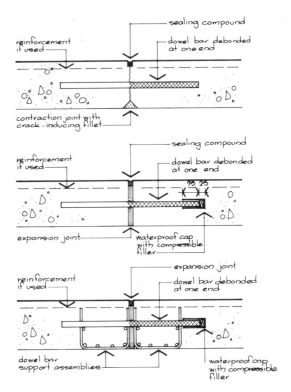

Figure 127 (top) contraction joint; (centre and bottom) expansion joints

The spreader should be operated at a speed consistent with the character of the material mix and thickness being laid, so as to produce a uniform density and surface texture. The material should be fed into the spreader at such a rate as to permit continuous laying as far as site conditions will permit.

The material should be rolled in a longitudinal direction as soon after laying as possible, but, without causing undue displacement of the material. The rolling should continue until all the roll marks on the surface have disappeared.

In the case of rolled asphalt, wearing courses containing 40% or less of coarse aggregate, 14 or 20 mm pre-coated chippings are spread on to the surface of the asphalt prior to rolling. They should be rolled into the surface while it is still plastic enough to cause some embedment of the chippings, this gives the running surface a roughened skid resistant texture.

Rigid construction

Once the sub-base has been prepared to the required tolerance the construction of the concrete slab can proceed. It is normal practice to use a separation membrane immediately under the slab to act as a slip layer, this membrane is invariably polythene sheeting.

In alternate bay construction used on smaller sites, side forms are set up for a number of alternate bays, leaving the intermediate slabs to be cast once the first bays have cured and hardened. The maximum bay length is governed by the joint spacing and would be approximately 13 m. Stop ends are placed in position at the correct joint spacing and the expansion and contraction joints with load transfer bars through them are set up prior to concreting see Figure 127.

Concrete is spread by hand over the slab and the reinforcement mesh is placed in position when the concrete has reached the required level. The top surface of the slab is compacted using a vibrating tamping beam. To enable tamping to be carried out the side forms are set to the finished surface level.

On larger road schemes; mechanised construction is employed. There are two basic methods in use in this country, slip form and fixed form paving.

In fixed form paving the spreading, compaction, finishing and curing of the concrete is carried out between side forms. These are normally steel and fixed to the sub-base by road pins, occasionally these side supports are a part of the permanent construction and take the form of concrete edge strips constructed in advance of the paving. Besides supporting the wet concrete these side forms also carry the rails on which the individual machines making up the concrete train run, Figure 128.

Slip form paving in which the plant has its own travelling side forms is the second method of concrete paving used in this country. It differs from the conventional concrete train in that all the operations of spreading, compaction and finishing are carried out within the length of a single machine. The travelling side forms provide edge support to the wet concrete only during the concreting operation. Therefore control on the concrete workability is essential. The machine is guided for line and level by sensors following wires set out accurately ahead of the paving operations, see Figure 129.

Figure 128   Concrete paving train
(Cement & Concrete Association)

Figure 129   Slip form paver
(Cement & Concrete Association)

18.8 Movement joints

In both methods of mechanised pavement construction the expansion and contraction joints are formed either in advance of the paving or by mechanical placing during the paving operation.

The purpose of the joint is to accommodate movement either arising from initial shrinkage or thermal movement of the slab or where the slab abuts an existing carriageway. Load transfer is achieved by means of mild steel dowel bars passing through the transverse joint.

Further reading

Readers who require a more in depth knowledge of some of the objectives in topic area E may find the following reading list useful.

E.16 Bartlett, Richard, *Design in Metric Sewerage*, Applied Science Publishers Ltd (1970).

Woolley, Leslie, *Drainage details in SI metric*, Northwood Publications Ltd (1971).

Exeritt, L. B., *Public health engineering practice; Vol. 2: Sewerage and sewage disposal*, MacDonald and Evans (1972).

Safety in sewers training manual, Local Government Training Board (1971).

E.18 Armeson, J. H., *Roadwork technology; Vol. 1, 2 and 3*, Iliffe Books (out of print) (1967).

O'Faherty, C. A., *Highway engineering; Vol. 2*, Edward Arnold Ltd (1974).

Ashworth, R., *Highway engineering*, Heinemann Educational Books Ltd (1972).

Croney, D., *Design and performance of road pavements*, HMSO (1977).

Walker, B. J. and Beadle, D., *Mechanised construction of concrete roads*, Cement and Concrete Association (1975).

Appendix – Road Note 29

EXAMPLE OF FLEXIBLE CONSTRUCTION

A flexible pavement is to be designed for an estate road for which the existing conditions exist.

(i) The subgrade is silty clay with a liquid limit of 50% and a plastic limit of 20%.
(ii) Assume a traffic growth of 4% per year, a design life of thirty years, the estimated traffic flow of commercial vehicles at the time of construction is six per day in each direction.

Traffic
Table 1. RN 29 places this type of road in category 1 that is up to ten commercial vehicles per day in each direction.

Figure 5. RN 29 indicates a cumulative number of commercial vehicles on each lane of about 0.3 million to be carried during its design life.

Table 2. In RN 29 the estimated number of standard axles per commercial vehicle is 0.45 × 0.3 million vehicles giving a total of 0.135 million over the design period.

Subgrade
The subgrade is silty clay with a Plasticity index of 30% (50%–20%) and since the water table is unlikely to rise nearer than within 1.0 meter of formation level the estimated CBR for the subgrade is 5% (Table 3 RN 29).

Sub-base
The plasticity index of the soil is 30% and the low water table level means that the possibility of frost action in the sub-grade can be discounted. There is no need for additional material to provide frost cover for the subgrade.

Figure 6. RN 29. CBR valve 5% gives a sub-base thickness of 180 mm.

Roadbase
Which may be wetmix or dry-bound macadam, cement bound materials. Dense coated macadam or rolled asphalt, will vary in thickness from 70 mm for rolled asphalt to 120 mm for wetmix and dry bound macadam. (Figures 7 to 10 RN 29).

Surfacing
See Table 4 column 4 RN 29 for alternatives, but 60 mm thickness is the design minimum for these particular circumstances.

EXAMPLE OF RIGID CONSTRUCTION

Using the same set of circumstances for the design of a concrete pavement the following steps are followed.

Table 5. RN 29 a CBR of 5% places it in the normal category and requires a minimum thickness of 80 mm for the sub-base. If construction traffic (loaded lorries) are to use the prepared sub-base then an additional 80 mm should be laid if it is considered that there is a risk of damage to the subgrade.

Figure 11. RN 29 shows that for a reinforced slab the thickness would be 135 mm rounded up to the nearest 10 mm. An unreinforced slab would require 150 mm thickness of concrete.

Figure 12. RN 29 requires that the weight of reinforcement would be slightly in excess of 2 kg/m^2, therefore a standard long mesh of 2.61 kg/m^2 would be specified.

Figure 13. RN 29 for reinforced slabs of the above weight of mesh a joint spacing of 16.5 m would be used. For winter construction every third joint would be an expansion joint, but for construction during summer all joints would be contraction. For unreinforced concrete construction the spacing of contraction joints would be 5.0 m. Expansion joints would only be included if the pavement were to be laid during the winter period, then spacing of the joints would be every 40.0 m.

Note. Since the publication of RN 29 (third edition) a Technical Memorandum number H6/78 has been published by the Department of Transport which revises the method of assessment of commercial traffic used as a basis for design. At the same time changes have been proposed to the lower layers of pavement to provide an improved foundation.

Readers should refer to the appropriate tables in RN 29 if the examples of flexible and rigid construction are to be fully appreciated.